Neil Kitching is an enigma. A geography graduate from Scotland, who qualified as an accountant in the private sector then worked as an auditor in the public sector. A burning desire to fulfil unmet dreams led to a mid-life career change to work in Sustainable Development policy. He now works as an Energy Specialist for a Scottish public agency supporting businesses. His role is to support the business environment for companies interested in water technology and renewable heat solutions - but Carbon Choices is his own work, independent of his employer.

Carbon Choices arose from a frustration that people are still unaware of climate change facts and its implications. Incorporating nature loss into the book is ambitious, but necessary, as nature is integral to climate change and wildlife is fragile. Neil welcomes constructive comments for a possible future edition, and can be contacted at carbonchoices@gmail.com

Please view his website, www.carbonchoices.uk, for further testimonials and information on where to buy Carbon Choices. Alternatively, to keep up to date, please follow him on social media.

www.facebook.com/carbonchoices

@carbonchoicesuk (twitter) @carbonchoices (instagram)

Front cover: sunset over the Okavango Delta, Botswana, 2019. It is the largest inland delta in the world, a wildlife haven with a strong community advocating nature protection, tourism and social development.

Praise for Carbon Choices

Dedication

To the fading memory of snow in our garden

Acknowledgements

To my parents and siblings who encouraged me to enjoy nature, explore places and to ask questions.

To my wife, Kay Kitching, who supported me through a mid-life career change, then gave me the space to write this book.

To good friends who kindly volunteered to review. Five core reviewers followed my progress chapter by chapter with extremely valuable comments: George Tarvit, Dr Gregor McDonald, Rupert Parkinson, Dr Sam Curran and my son Calum Kitching.

Special thanks to Alan Speedie, consultant; Chris Wood-Gee, former chair of Sustainable Scotland Network; friend Steven Scott, my sister Dawn Waitt and my father Brian Kitching who sense-checked the whole book. And to other reviewers, including my son Gavin Kitching and friends Josie Stansfield and Sue Walker. To Tom Smith who proofread the book.

To friends, work colleagues and other specialist reviewers of individual subject chapters who commented in their spare time; Aileen Lamb, Catriona Patterson, David Pickett, Emma Waddell, Erik MacEachern, Ewan Mearns, Ken Maxwell, Kirsty Platt, Neil Ferguson, Niall Williamson, Pete Ritchie, Scot Mathieson and Stuart Adair.

Alex Hill, Met Office, Chief Government Advisor (retired)

Andrew Smith, Director, Greenbackers Investment Capital

Dr Andy Kerr, Director (UK & Ireland), EIT Climate-KIC

Barry Carruthers, Head of Innovation & Sustainability, Scottish Power

Ben Lynam, Head of Communications, The Travel Foundation

Clive Mitchell, Open University and Outcome Manager, NatureScot

Colin Webster, Learning Content Manager, Ellen MacArthur Foundation

Dave Pearson, Group Sustainable Development Director, Star Refrigeration

Ed Craig, Director Innovation ECCI, Dean Low Carbon College

Eric McRory, Chief Economist Scottish Environment Protection Agency

Francesco Sindico, Co-Director Strathclyde Centre for Environmental Law and Governance

Gordon Waitt, Lecturer at University of Wollongong

Jim Panton, Managing Director, Panton McLeod

Konstantinos Kontis, Professor of Aerospace Engineering, Glasgow University

Dr Martin Valenti, former vice-chair of Scotland's 2020 Climate Group

Mike Barry, Director, Mikebarryeco

Mike Robinson, Chief Executive, Royal Scottish Geographical Society

Neil Glasser, Professor of Physical Geography, Aberystwyth University

Osbert Lancaster, Director, Realise Earth

Dr Rowan Ellis, James Hutton Institute

Professor Sean Smith, Chair of Future Construction, University of Edinburgh

A number of other people assisted including Sue Mitchell on my writing style, Tom Smith developed my website and was my first 'follower' on twitter, Sandy Fleming who advised me on Facebook, Ingrid Roughead spurred me to think about marketing, David Hamilton on self-publishing and my son Alasdair Kitching who did further research on how to self-publish.

To my employer, who believed in me, and enabled my career change from accountant to Sustainable Development Policy then to Energy and Water specialist.

And finally, to the information gleaned from the internet without which this book would not be possible.

Contents

Section 3: Solutions - Applying the Building Blocks to our Lives

Foreword

I was fortunate to be brought up in a family that enjoyed visiting the Scottish countryside and observing nature. My parents gave me the freedom to explore mountains, woodlands and rivers and to cycle on roads from a young age. This led to my interest in geography. However, in my lifetime the climate has noticeably changed. Winters are warmer with less snow, and increased heavy rainfall causes floods and landslips in places where they did not occur before. In my childhood there were more flying insects; car windscreens were regularly splattered with dead insects and you could not leave a window open in the evening as insects would be attracted to the light. Unfortunately, whilst many insects have declined, milder winters have helped to increase others such as blood sucking ticks and the diseases that they carry.

Our society has also changed. Road traffic has increased dramatically, towns and cities have grown and green spaces have shrunk. I was raised in a house on the edge of Perth, a large town between the Lowlands and the Highlands in Scotland, now surrounded by sprawling housing estates. Eating out has become a normal activity, rather than an occasional treat. And, at primary school I remember only one classmate who flew to Tenerife each year for a family holiday - now foreign travel is commonplace for many and an expectation for some.

I have witnessed the explosive growth in urban population. As a student I visited New Delhi in 1984, population seven million; and returned on a work trip in 2018, population nearly 30 million. In 1984, the streets were crammed with a crazy mix of lorries, cows, oxen dragging over-loaded carts, bicycle rickshaws and people carrying heavy loads. This time the streets were equally busy but full of polluting lorries, private cars and dirty diesel rickshaws; with cows and pedestrians adding to the mix. A blanket of air pollution hung over the city. Meanwhile, the total human population has grown in the five decades of my life from 3.4 to 7.7 billion.

Humans have caused many adverse environmental impacts; many have worsened in my lifetime. Intensive agriculture overexploits the thin layer of fertile soil that we all depend upon, depleted of nutrients and blown or washed away. Downstream, estuaries fill with nutrients from sewage and the run-off from agriculture. These nutrients enable algae to multiply rapidly which depletes the supply of oxygen in the water causing fish to suffocate. Our rivers and oceans are full of plastic, and litter is a problem on land especially in low income countries that cannot afford the infrastructure to collect or recycle the waste. Species rich tropical forests are still being cut down leading to a loss of wildlife. Acid rain forms downwind from coal power stations, whilst air pollution cuts the life expectancy of millions of people.

It is not all doom and gloom. There have been many environmental improvements. The insecticide DDT was a new wonder chemical that killed mosquitoes, reducing the incidence of malaria. Unfortunately, DDT became concentrated further up the food chain and affected predators like birds of prey. It caused their egg shells to thin and to break easily, and led to several bird species becoming near extinct. One of the early environmental campaigners, the American Rachel Carson, made this connection, and led a successful campaign to ban DDT in most places. Within Scotland, there have been many environmental improvements such as removing lead from petrol, salmon have returned to the formerly heavily polluted River Clyde, new cycle paths are being built, and more native deciduous trees planted. Increased efficiency has led to a steady fall in the total energy consumed in the UK since 2005. Recently, after decades of indifference, and following a campaign by Sir David Attenborough, the public has finally taken notice of the impact of plastic on our environment. Things change, just not quickly enough.

****

For the last ten years I have worked on climate policy and company innovation in Scotland. Scotland is a small, but affluent, country of five million people, one of four countries that make up the United Kingdom (UK). The industrial revolution began in the UK in 1776 when a Scotsman, James Watt, developed a more efficient steam engine that produced more power and burnt less coal. Initially it was used to pump water out of mines, and this reduced the cost of extracting coal. The number of steam engines then grew quickly and spread across the world. Global emissions of carbon dioxide accelerated from the 1770's onward.

Moving forward to November 2021, Glasgow, the largest city in Scotland, hosts the United Nations climate change conference (COP26) – a conference crucial to our future. Hopefully, the climate change conference will build on and strengthen the existing climate treaty, the Paris Agreement, which does not contain strong enough commitments from every country to prevent dangerous climate change causing catastrophic impact on humans and wildlife. The Covid-19 outbreak delayed this conference from November 2020. This pandemic has shaken global society and economy, resulting in changes which previously would have been politically unacceptable. I have assumed that many, but not all, of these changes will be temporary and that country-wide lockdowns will have passed by the time you read this. A chapter towards the end focuses on the lessons learnt from the pandemic.

Scotland is an excellent place to host this conference. It has a devolved parliament with responsibility for many areas of climate policy. The Scottish Government has strongly pushed action on climate change since 2006. In 2009, two headline targets were set for 2020: to produce more renewable electricity than total electricity demand and to reduce all climate warming emissions by 42%. A new Climate Change Act incorporated this latter target, along with a long-term commitment to reduce carbon emissions by 80% by 2050. At the time these seemed like bold and ambitious targets. A well-known professional engineering institute argued that the 100% renewable target was set by politicians, not engineers, and that it could not be sensibly achieved. In fact, Scotland is likely to achieve its 2020 renewable electricity target and total emissions have already fallen by over 50%.

In 2019 the Scottish Government announced a climate 'emergency' and set tougher targets. They tightened the existing carbon reduction target to a 100% 'net zero' target by 2045. This includes Scotland's share of emissions from international shipping and aviation. The ambition is for widespread charging infrastructure to be in place for electric cars by 2032, new buildings are to be heated by renewable or other zero carbon sources from 2024, all trains are to be emission free by 2035, and flights within the Highlands are to be low carbon by 2040. In addition, Scottish Water, the state-owned water utility, has set itself an ambitious target to be net zero by 2040. Ambitious because this target covers emissions from its own operations; and all emissions from its significant capital construction programme – all their suppliers who extract raw materials, produce concrete and steel, transport materials and construct infrastructure.

Scotland is a small country, trying to set a good example for other countries to follow. But this book is not just about Scotland. I use international examples of good and bad practice throughout.

****

There are many books on climate change. Some focus on the science; others tell us the steps that individuals can take to take to reduce their impact. This book will not repeat everything that there is to know about climate change. Instead it will provide enough information to help you to understand the science of climate change, and to be better informed to make reasoned choices about your impact on the environment. I call these "carbon choices". This book takes a holistic view covering an eclectic mix of science, ecology, engineering, psychology, economics, politics and geography. It considers the twin environmental crises rooted in unsustainable development that humans face; that of climate change and loss of nature. The two are intertwined. Healthy forests, grasslands, savannas, wetlands, peatlands, soils, mangroves, sea grasses and coral reefs all absorb and store carbon dioxide. Conversely, human impacts and destruction of these ecosystems can often release this stored carbon leading to more emissions of greenhouse gases.

I chose the title, "Carbon Choices" because this book will provide the information that you need to make better choices to reduce your carbon emissions and to think about your impact on nature. The sub-title "common-sense solutions" highlights that many of the solutions to climate change and destruction of nature are based on informed common sense.

Section 1 *introduces* carbon dioxide, climate change and the destruction of nature in a comprehensive but non-scientific way and summarises the impacts of climate change and how we have descended into an existential environmental crisis.

Section 2 introduces ten *building blocks* that will lay the foundations to enable us to make better choices for the planet.

Section 3 introduces *five principles* that will guide individual consumers, businesses and governments to make better carbon and environmental choices. I apply these principles to aspects of our day to day lives to illustrate what we can do if we choose to do so.

The book's conclusion includes a green action plan for governments, businesses and individuals.

I use as few technical terms as possible, but some are unavoidable. 'Carbon' is used as shorthand for carbon dioxide and other greenhouse gases that contribute to a warmer atmosphere – often called global warming or global heating, but perhaps more accurately described as climate change as not all areas will necessarily warm. A 'carbon footprint' is the amount of carbon emitted from the production and use of goods or services. A 'low carbon economy' is a society and economy that does not emit much carbon. 'Net zero' is where any remaining emissions are offset by an equal amount of carbon being permanently removed from the atmosphere. 'Wildlife' is used to describe all living species, including plants, animals, birds, and fungus; 'nature' has a wider definition that includes wildlife but also landscapes and soils.

****

Given our intelligence, education and science, you might think that protecting our planet for humans and nature should not be that difficult. To do so, we need a wide-ranging mix of actions; a recipe to make a better future. The ingredients, explored further in Section 2, are:

- plan for the long-term
- directed taxes and subsidies
- sensible regulations
- good design
- targeted innovation and investment
- education and training
- behavioural change and peer pressure
- a commitment to work together for a better future

In short, we need an approach based on common-sense.

Think long-term. Get the price right. Incentivise good behaviour. Ban the bad.

To add flavour to this recipe, sprinkle on some community action, charity and philanthropy, and consider how one thing affects another and be fair. Tear up vested interests. Businesses should work with government to promote change, instead of wasting time and effort resisting change that is clearly necessary.

Currently the world has a mix of contradictory policies and actions. Climate change deniers have fought hard to block or delay progress to cut carbon emissions. Human society is not taking climate change or the loss of nature seriously. But we can if we want to. It is not about hope or capability. It is whether we as humans choose to work together for a better future. Every time we have the opportunity, we need to make the right choice, and use these choices to influence each other, businesses and our politicians. One seemingly insignificant change can be a catalyst for so much more.

Section One:
Understanding Carbon and its Impacts

Chapter 1:
Why is it so Difficult?

Tackling climate change is proving to be difficult. To tackle a problem, humans need to understand, and believe in the science, and then overcome any structural, economic, social and political barriers. Preventing climate change is the ultimate difficult problem due to its complex scientific, political and economic challenges and because its impacts are not instantaneous. It would be hard for writers of science fiction to dream up a more insidious plot to degrade the environment of an entire planet and make it less habitable for its dominant species and for its wildlife.

Human climate change is caused by our use of fossil fuels, consumption of resources, agriculture and impact on land-use. These activities produce or release greenhouse gases which trap heat in the atmosphere. Nitrogen and oxygen make up most of our atmosphere but are not greenhouse gases. The biggest contributors to human induced climate change are carbon dioxide, methane then nitrous oxide. They all have more complex molecules than nitrogen and oxygen that vibrate when heat from the surface radiates up to them. They absorb some of this heat and radiate it out in all directions, with some returning towards the Earth's surface. This is the greenhouse effect that makes the Earth habitable. Without it our oceans would freeze. But, like wearing a jacket on a hot day, humans are increasing the volume of greenhouse gases in the atmosphere which causes it to warm beyond what is normal or comfortable.

Carbon dioxide from human activity comes from burning fossil fuels and from the decay of vegetation such as trees when cut down. It is an invisible, odourless, colourless gas. Its effects are global and long term. Unlike most environmental actions, there is no immediate benefit to an individual or a country from reducing their emissions. It is a strange thought that we are rapidly releasing gases which were captured slowly over millions of years by natural processes; in the case of coal, most was formed from decaying dead plant matter in the Carboniferous period around 300 million years ago.

Methane is produced from natural sources such as degraded peatlands, volcanoes and wetlands where vegetation decomposes without the presence of oxygen. But more than half of the emissions are from human activity. The main source is agriculture; dominated by cattle and sheep burping but also from flooded rice paddy fields. The second main source is leaks - from coal mines, oil and gas wells, pipelines and from burning natural gas to produce heat or electricity. It is also emitted from rotting food waste such as in landfill sites.

Nitrous oxide, commonly known as laughing gas, is produced from many sources including applying fertiliser and vehicle exhausts. Fertiliser is produced by converting natural gas into hydrogen and combining it with nitrogen in an energy intensive process to create ammonia. 1% of global carbon emissions are from manufacturing fertiliser and a further 1.5% from its use.

Measuring carbon dioxide emitted from burning fossil fuels is relatively straightforward. It is far more difficult to measure some other sources of carbon dioxide and other gases. Measuring emissions of methane and nitrous oxide from fertiliser, land-use change, and burping livestock is difficult even in high income countries. Imagine how much harder it is to do so in low income countries with less money and fewer resources available to collect reliable data. In addition, factories may illegally under-report emissions of banned climate change and ozone destroying gases such as chlorofluorocarbons (CFCs).

Historically there is a close correlation between economic growth and carbon emissions. Attempts to reduce our emissions can challenge our societal aspirations of a better life for us and our children. Until Covid-19, the only significant drop in global emissions occurred in the aftermath of the global recession of 2010. This reinforces some people's belief that our economy cannot afford to cut its carbon emissions. However, individual countries with strong climate policies, such as Sweden and the UK, have managed to decouple the historic correlation between economic growth and emissions. They have grown their economies and cut emissions at the same time. Some think that a global recession, or a pandemic, are the only ways to cut our carbon emissions, but recessions delay investment in much needed clean technology and the infrastructure which is needed for us to lead satisfying low carbon lifestyles.

Another difficulty is that it is not clear who is responsible to control the international emissions from shipping goods and from international passenger flights. For shipping, is it the ship-owner, the country that registers the ship, the departure or arrival port, the country that produced the goods or where the goods are consumed? For aviation is it the country from or to the plane flies, or is it the nationality of the passengers? Even worse, the adverse climate impact of aviation is more serious and complicated than just the carbon dioxide emissions from burning aviation fuel. Planes also emit nitrous oxide, and water vapour which can form contrails. These can spread to create high level clouds. During the day, these cause a small cooling effect as they block some incoming solar radiation; at night they prevent heat escaping and keep nights significantly warmer. Therefore, aviation decreases the range in temperature between the maximum during the day and the minimum at night. Interestingly, this was confirmed after 9/11 when the USA cancelled all flights for three days. The range in daily temperature in populated areas increased (back to its 'normal').

Humans are not good at self-control and dealing with slow developing long-term issues. We evolved to respond to immediate external threats – the flight or fight reflex. Examples where many of us take a short-term approach, that might prove to be ultimately harmful to us, are our attitudes to skin cancer from sunburn, long-term tooth decay from sugary drinks, obesity from over eating and lung disease from smoking. Many people anxiously wait for their next monthly or weekly pay; or, in some countries, are anxious where their next meal will come from. Even in the wealthy UK, the number of people that rely on free food banks is increasing so perhaps it is not surprisingly they are less likely to spend time thinking about the long-term. Humans also tend to think that 'it won't happen to me', 'I am too busy', 'I will deal with that later' or that climate change is someone else's responsibility to sort. To tackle climate change, we will need to reverse this thinking and tackle long-term challenges to benefit society and nature.

Some even deny that climate change exists, deny aspects of it or argue for a delay in action to cut carbon emissions. The range of arguments can be summarised as follows:

- Climate change is not happening, or it has always happened and is natural.
- It is better to 'develop' now to pull people out of poverty; we can adapt to climate change and future technology will solve the problem.
- Reducing our emissions will reduce economic growth, undermine progress to alleviate poverty in low income countries or will disproportionately impact certain sectors or sections of society.
- It is pointless to act because countries like China cause more impact than us, and population growth will increase emissions elsewhere in any case.

However, the basic science on human induced climate change is clear. On the economic argument, some experts argue that future generations will be wealthier, and will be able to tackle any problems that humans create. Others, such as Lord Stern, take the opposite view that, like an insurance policy, acting now will be more cost effective. Some problems grow exponentially, such as the spread of invasive species; some changes are permanent such as loss of a species; and tipping points may accelerate climate change. The impacts of climate change will cut future economic growth and the more carbon we emit the more climate change will accelerate. Delay is not a sensible course of action.

Those that wish to delay action will focus on social justice arguments or push for actions that are relatively trivial – a distraction rather than the transformational changes that are needed. Examples include proposing to build gas power stations rather than coal or arguing against a carbon tax on aviation saying that it will hurt hard working families. In fact, using gas will simply prolong our dependence on fossil fuels and any tax on aviation will impact wealthy people more than those who are less well off.

For a long time, politicians and the public assumed that the price of resources, including oil and gas would increase steadily as access to these resources ran out or became difficult or expensive to extract. These price rises would naturally restrict our consumption and therefore limit our potential damage to the environment. But these arguments have almost always been disproved. New technology such as hydraulic fracking has enabled companies to discover and extract more oil and gas cost effectively. In any case, if prices rise, the market adjusts. When oil prices rose dramatically in the 1970s more effort was spent developing fuel-efficient vehicles to reduce demand, oil power stations were converted to gas, and more money was spent exploring for new sources of oil such as in the North Sea. It looks like we cannot rely on shortages of key commodities and consequent price rises to limit the adverse environmental impact of humans.

Politicians cannot agree on the best approach to tackle climate change. On the left politicians argue for social justice often linking climate change and the environment to poverty and injustice. Big polluters should pay. There should be more regulation on business. On the right, politicians dislike regulations and see climate change policies as working against growth and business and as an attack on 'freedom'. Meanwhile, trade unions worry about the impact of new regulations on the jobs and futures of their members.

But humans have successfully tackled other environmental challenges - and we can do this again if we have the political will. Through international action we have reduced the impacts from acid rain, the hole in the ozone layer, air pollution, lead in petrol, and pollution from sewage to name a few.

Some of these challenges are local and these have been the easiest to solve. Lead is a neurotoxin which causes brain and nerve damage. The UK phased out lead in petrol, firstly by imposing a tax differential between the price of leaded and unleaded petrol, followed later by imposing a ban. Unfortunately, lead from sources such as unregulated recycling of car batteries, is still an issue in many low income countries.

Acid rain is a regional issue. Downwind from coal fired power stations, sulphur dioxide reacts with water to create sulphuric acid which adversely impacts lakes, wildlife and can even damage building materials. The European Union (EU) successfully solved this over several decades by introducing regulations on coal fired power stations. In the 1970s it seemed like an impossibly expensive problem; now it is no longer a concern in Europe, although it is still an issue in countries like China.

Chlorofluorocarbons (CFCs) were a genuine global issue as emissions from all countries were mixing in the atmosphere to destroy ozone in polar regions. The ozone layer protects us from much of the harmful ultra-violet radiation from the sun. Following the discovery of the ozone hole over the Antarctic, and to many commentators' surprise, governments quickly enacted the Montreal Treaty. It is successfully, albeit gradually eliminating this problem. The world was shocked by this new environmental issue, and it was a small and manageable number of chemical companies in a small number of countries who manufactured CFC's.

So, there are grounds for optimism if countries choose to work together. Unfortunately, a wide range of activities which are essential to our welfare cause climate change. In fact, our economic growth has depended on access to cheap fossil fuels and there are so many strong vested interests at stake.

Given politics and national interests, global agreement on climate change is bound to be difficult and complex. There are economic, political, environmental and social justice issues to consider. Why would Saudi Arabia agree to stop pumping oil; and why would a country stop burning their cheap coal when they won't directly benefit? Should countries with large historical emissions cut back more and faster? Can we even measure our emissions, particularly from land-use, with any degree of accuracy? Will countries who are less impacted by climate change want to act as fast as those, such as semi-arid countries or low-lying island nations, who will be most impacted? Should we allow low income countries to increase their emissions to raise the standard of living of their people? Should high income countries fund low income countries to help them to reduce their emissions, or to adapt to climate change? What if the donor or recipient country is rife with corruption?

Ultimately, climate change is a difficult problem that requires action by international, national and local politicians, businesses, communities and individuals. The best balance of action between these is not clear. All need to act but radical change is often led by young people. And, why is it so difficult to solve climate change? Imagine playing Cluedo, the game where you try to deduce how someone was murdered, where the murder was committed and who the murderer was. But, imagine playing the game in the dark. The murder weapon is invisible (carbon dioxide), the location does not matter (carbon dioxide emitted anywhere spreads around the world) and the players (countries) are all complicit in the murder and not cooperating with the investigation.

Chapter 2:
The Perfect Storm

Even with all the disasters you hear on the news, the majority of us have not yet been directly affected by a natural disaster - a hurricane, major flood, drought or devastating fire – nor needed emergency support to obtain shelter, food or to access healthcare. Most of us have been inconvenienced in minor ways: a lost tile from the roof, a flooded garden, a road blocked by a fallen tree, but never anything that has been a serious practical or financial challenge. Natural disasters are things that we hear about on the news. They affect other people and in other places that are often remote and exotic. Of course, nowadays many of these weather-related natural disasters are exacerbated by climate change. We used to call them "acts of God". We could more accurately call them Human Amplified Natural Disasters, giving the acronym HAND.

Climate change will disrupt us all. The environmental changes will affect us personally. They will impact you, your family, your property and society. Climate change will have a disproportionate impact on the poor, the disadvantaged and minorities. It will have a major impact on a country's wealth and on your own well-being and security through impacts on global supply chains, food prices and cost and availability of insurance. Climate change, often played out through water shortages, can function as an additional stress on the environment and society leading to conflict and an increased likelihood of authoritarian government. Stresses can tip people over a threshold that encourages or forces them to migrate, creating millions of global refugees. Climate change will devastate wildlife. A United Nations (UN) report estimated that a million species are at risk of extinction. Climate change will herald a shift in global ecosystems to new locations, but with less diversity of wildlife.

There is little good news when you consider the scientific facts around climate change. Some areas might receive a boost in agricultural productivity from warmer temperatures, and for those in cooler climates there will be a reduction in heating bills. But on balance the change is very bad. Our modern human society developed during a period of relative climate stability over the last 10,000 years and any change from this will cause major disruption.

Human induced climate change is already shifting our climatic zones towards the poles. Vegetation and wildlife, and our agriculture and human populations, are dependent on the stability of these climatic zones. Any shift will inevitably cause disruption. Before human interference, wildlife could adapt and move if climate change was slow enough. But now human induced climate change is too rapid for many species to adapt, and barriers created by humans, such as motorways or areas of intensive agriculture, often block wildlife migration. Our precious nature reserves and the wildlife within them will become 'trapped' in increasingly unsuitable climatic zones.

Human population density is closely correlated to favourable climatic zones and soils, therefore any shift in the climate will result in hardship and mass migration. Unless there is artificial air conditioning and imported food, it is difficult for humans to survive in a climate where the average air temperature exceeds 29°C. Today, only a few in the Sahara and Middle East endure these conditions, but the numbers affected will swell rapidly with each degree increase in global temperature. Many populous countries, like Bangladesh that exists on a river delta, are particularly vulnerable to flooding mainly from rising sea level. This will cause enormous social difficulties as there are physical, religious or political barriers to migration.

Climate change is not just about an increase in temperature averaged over a long period of time. Many of its health impacts result from heat waves and extreme heat, often exacerbated by drought. If human bodies have no time to recover from warm and humid temperatures at night, then this can lead to heat stroke and deaths even in temperate regions like Europe. Similarly, for nature, extreme heat stress can be the catalyst for mass deaths and even extinction of species. Bleaching of coral is the best-known example, but it can impact all types of wildlife, from insects to birds, wherever the conditions are too extreme for the temperature range that animals or plants evolved in. In 2018 in Australia, heat stress overcame endangered flying foxes and over a few days 23,000 fell dead out of trees, whilst trees themselves are susceptible to the twin threats of heat stress and drought.

Water is the primary medium through which we will feel the effects of climate change. Weather patterns will shift causing droughts in areas that previously rarely experienced them, and more intense droughts in already dry areas. Warmer temperatures increase evaporation from the soil worsening the impact of drought. Conversely an increase in evaporation over oceans puts more water vapour, and consequently energy, into the atmosphere leading to torrential rainfall and floods. Furthermore, bacteria grow faster in warmer water resulting in the spread of algae blooms that reduce oxygen in the water, in some cases causing fish to slowly suffocate.

Water is required for life. There is a lot of fresh water on our planet, but it is not always in the right place, of the right quality, at the right time. Human activity already affects the water balance. Glacial retreat in the Himalayas risks future summer drought in the populous regions downstream. Diverting rivers to irrigate cotton fields has destroyed the Aral Sea in Central Asia. In parts of India groundwater is falling due to over exploitation from pumps and boreholes for agriculture. Similarly, diverting water for agriculture has wiped out migratory fish from the Colorado in the USA as the river no longer reaches the sea. Manila, capital of the Philippines, and Jakarta, the largest city in Indonesia, are both sinking due to over exploitation of groundwater. The wildlife haven of the Okavango Delta in Botswana and people living on the edge of the Sahara Desert are already at risk from drought. Climate change will make all these impacts worse through rising sea-level and changes in precipitation patterns.

Many religions and societies have flood myths within their ancient beliefs. Maybe this is because flooding is so traumatic to people and societies. It is not just a matter of drying a property out. Often floodwater contains raw sewage which can lead to the risk of disease. Flood victims are typically displaced for a long period of time, leading to depression and mental health issues. Then they cannot renew their household or business insurance creating uncertainty and financial stress.

You might think that ice sheets are stable, but they are not. They are susceptible to minor shifts in the Earth's orbit which change the distribution of energy that reaches the Earth's surface. This has driven past Ice Ages where ice sheets repeatedly grew in Europe and North America, expanded then collapsed. Due to so much water being bound up in ice, at the height of the last Ice Age, sea level was an incredible 120 metres lower.

Today, warming is happening fastest in the polar regions and sea level is rising due to melting ice from Antarctica, Greenland and valley glaciers. Moreover warm water expands, takes up more space and pushes up sea level even more. Clearly this increases the frequency of coastal floods and the risk of permanent displacement of people and wildlife.

The precise future of sea level is hard to predict because ice sheets respond slowly to changes in climate, taking a long time to reach a new equilibrium. But it is sobering to know that sea level will continue to rise for hundreds of years even if we stop emitting greenhouse gases today.

If all the ice in Antarctica melted sea level would rise by a further 60 metres. Of more immediate concern is Greenland. Sea level would rise by seven metres if all its ice melted. Greenland is a relic of the last Ice Age, kept cold by its self-created altitude – its huge depth of ice built up over thousands of years. As a result, if it melted completely the ice sheet would not reform under the present climate. This ice sheet is now vulnerable as the climate in the Arctic is warming fast. On its surface, meltwater and black soot particles from pollution and fires darken its surface and absorb heat, whilst warm water is encroaching from the oceans to melt ice from underneath. I have visited the Jakobshavn glacier, one of the fastest moving glaciers in the world. It drains a large part of Greenland into a dramatic fjord full of icebergs. It looks like one of the most untouched, natural places in the world. But the calving front of the glacier into the sea has retreated 40km since 1851. So, this huge ice sheet is already melting, and this is accelerating. Even a two-metre rise in sea level would affect 630 million people living along coasts. And it looks like we cannot prevent this.

Global warming is also likely to lead to more intense cyclones and hurricanes, driven by warmer surface ocean temperatures. Gradually rising sea level will multiply their destruction, exposing more areas to storm surges. Such events do not just cause immediate damage; when saltwater encroaches inland it can change the soil chemistry and make the soil infertile to most agricultural crops. Rising sea level also threatens the drinking water intake for many low-lying Pacific islands and coastal cities and to their sewage outflows that use gravity to discharge to the sea.

In addition to saltwater encroachment along coasts, there is a further even more insidious threat from changes in ocean chemistry – ocean acidification. Life depends on oceans for many things, including algae and phytoplankton that produce much of our oxygen. But as atmospheric carbon dioxide increases, the oceans absorb more where it reacts with water molecules to form carbonic acid. Less alkaline (or more acidic) water dissolves and weakens calcium carbonate-based sea organisms such as seashells, crabs, corals and even fish bones. We are unwittingly undertaking a giant chemistry experiment with our oceans, and if we continue to make them less alkaline then this will cause local, then regional collapses in marine eco-systems. It is thought that changes in ocean chemistry instigated most previous mass extinctions in the geological record (except for the asteroid that wiped out the dinosaurs). An important conclusion is that we must reduce carbon dioxide in the atmosphere to reduce the oceans absorbing it. Any other attempt to reduce global temperatures, such as deliberately putting aerosols into the atmosphere to reflect incoming solar radiation, might succeed in reducing temperatures but would not protect the oceans that life depends on.

Warmer temperatures also lead to more evaporation creating more potential energy and convection in clouds. This increases the number of thunderstorms and lightning which produces nitrous oxide. Lightning strikes start around half of all wildfires. Because of the warmer temperatures and prolonged droughts, fire-fighters have stated that the recent wildfires in California and Australia are different than those in the past. They are hotter, spread faster and burn deeper into the soil. This causes long-term damage to vegetation and soils, unlike natural fast burning, lower temperature fires. Intense fires leave the soil bare and vulnerable to intense rainfall resulting in soil erosion and landslips. The ash can wash into rivers and harm fish. These devastating fires cause human fatalities, air pollution, and death and destruction of wildlife. The Australian bush fires of 2020 released more carbon dioxide into the atmosphere than Australia normally emits in an entire year. Peat fires are also increasing due to warmer temperatures that dry the soil. They can burn for months, as evidenced by recent fires in Indonesia and even in arctic Greenland releasing large quantities of carbon dioxide.

Warmer temperatures also threaten our health, enabling pathogens, pests and diseases to spread from the tropics towards the poles. Malaria and dengue fever are spreading out from the tropics, now affecting people who have no natural immunity. This spread of diseases will adversely affect the health of animals, crops and humans. One third of amphibians, such as frogs, are under threat from an infectious fungal disease which spread from Africa probably facilitated by climate change, whilst the mountain pine beetle is devastating forests in western North America. Its spread has been enabled by the reduction in hard winter frosts that used to limit the beetles' range, and tree death is accelerated by the stress the trees already face after years of below average rainfall.

****

In addition to all the individual problems that arise from climate change, there is an even bigger concern; that of climate tipping points. Like the fall of the Berlin Wall in 1989, change is not always slow and steady. It can start slowly, and then suddenly switch to a new equilibrium, very different from the original situation and potentially irreversible. Similarly, climate change may not always be slow, and sudden shifts may happen worsening the impact on humans and wildlife. For example, swathes of the Amazon rainforest could rapidly change to savannah grassland following random droughts, clear felling and fires. Without the trees there will be less evaporation from leaves leading permanently to less rainfall. Likewise, coral reefs are particularly vulnerable as they are sensitive to rising water temperature that causes the coral to 'bleach' and die. Around one quarter of all marine life is dependent on coral reefs and 250 million people depend on them for their livelihoods. The reefs protect low-lying islands from waves and storm surges. Once damaged by warm water they take decades to recover if at all. Scientists predict that most coral reefs, upon which so much marine life depends, will be severely damaged by a two-degree rise in ocean temperatures.

Many of these impacts will lead to 'positive feedback' effects that make the problem worse. But this is not 'positive' news at all. These feedbacks will result in a larger increase in temperature than might be expected from the direct effect of more greenhouse gases in the atmosphere. It is these feedbacks that amplify natural changes to the Earth's climate, shifting the world between Ice Ages and warmer inter-glacial periods, initiated from relatively small changes in the Earth's orbit around the sun.

A strong feedback that helps ice-sheets to grow and decline is because snow and ice are white which reflects most of the incoming solar radiation. If snow or ice covers the ground and sea then the surrounding area remains much colder than if it melts to be replaced by tundra, woodland or open ocean. In the Arctic, disappearing reflective snow and ice has already led to an increase in average temperature of several degrees centigrade, much greater than the average global increase. This increases melting resulting in a slow, long-term loss of ice and snow. Assuming current trends continue, scientists predict the Arctic Ocean will have ice-free summers before 2050.

The impact of changes in the Earth's orbit and reflection of solar radiation by snow are then reinforced by complex changes in the carbon cycle. As ice sheets grow, sea level falls enabling carbon based vegetation to expand, whilst the colder water can also dissolve more carbon dioxide and remove it from the atmosphere. This reduction makes the climate even cooler. These processes reverse as Ice Ages end, again initiated by small changes in the Earth's orbit.

It is not just change in surface snow and ice that affect our climate. Permafrost is frozen soil that covers one quarter of the land area in the northern hemisphere in Siberia, Canada and Alaska. It is already beginning to thaw. The once frozen organic matter in the soil decomposes, emits methane and carbon dioxide and adds to further global warming. And as Arctic temperatures rise, trees and shrubs colonise northwards. Trees shed snow more quickly than the tundra that it is replaces. The land changes from white to green resulting in less reflection of solar radiation and a further rise in temperature which in turn thaws more permafrost and releases more methane.

Then there is a conveyor belt of ocean currents that transport heat towards the poles. For example, the Gulf Steam carries warm water from the tropics to the Arctic, and in the process, it keeps north-west Europe comparatively warm. There is a risk that this current could slow down or even collapse. Evaporation drives this current, leaving a surface of salty water in the Arctic Ocean, which is denser than freshwater and therefore sinks to the deep ocean. But melting ice from Greenland is adding more freshwater to the Arctic Ocean. This dilutes the salty water making it less dense and this could reduce the power of the engine that drives this entire ocean current. Ironically, this could make parts of north-west Europe, including the UK, cooler and wetter than it would otherwise be. Moreover, this sinking water in the Arctic removes carbon dioxide from the atmosphere and buries it in the deep oceans for hundreds of years. If this mechanism slows then we will lose a significant natural method to drawdown carbon.

Another natural carbon store is vegetation and forests that soak up some of the excess carbon dioxide that we emit into the atmosphere. Fire and drought weaken or destroy trees and can adversely impact soils and emit greenhouse gases. If fires continue their upward trajectory then another natural carbon store may be lost. And in future tropical wetlands may emit more gases as warmer conditions increase microbial activity.

A further issue is clouds. They hold a lot of water vapour, a natural greenhouse gas. Clouds have a complicated but significant impact on the climate – they reflect incoming solar radiation during the day but also function as a blanket reducing heat loss from the surface at night. Some recent models of climate change indicate that there is a risk of significant changes in cloud cover in a warming planet. Low-level stratus clouds may become thinner or burn off completely allowing more solar radiation to reach the surface. The science around this is still uncertain but this is an impact that could increase temperatures even more than current climate projections.

Climate change will clearly impact human society. Extreme weather events are local, some climate change impacts such as droughts are regional, and sea level rise is global. Adverse weather in key crop growing regions can lead to global increases in the price of staple food crops, whilst migration from one country can impact other regions. For example, some blame the collapse of society in Syria on a severe drought which forced people to move from rural areas to the cities which increased the ethnic mix and tension. A collapse in the Russian wheat crop increased prices in Syria triggering ethnic conflict.

Not all adverse environmental impacts are due to climate change, but it can act as an additional stress that pushes the environment or people over a threshold. Droughts caused by climate change add to existing human overexploitation of water resources. Landslips are caused by a combination of more intense rainfall, impacting on bare soils exposed by humans removing trees from steep hillsides. Whilst non-native species are mostly spread by human activity, climate change can exacerbate this, helping some species to spread to new habitats where they outcompete native wildlife.

To summarise, climate change is not good, and is potentially disastrous, for us and wildlife. Many of its effects, such as sea level rise will be irreversible except on a geological timescale. That is why we all need to act now and act fast.

Chapter 3:
Carbon Literacy

At the 2015 UN climate change conference in Paris almost every country committed to the 'Paris Agreement'. Each country agreed to publish plans to reduce their carbon emissions, or in the case of most fast developing countries, to reduce their previously expected increase in emissions. This agreement set a target of a maximum increase in average global temperature of 2°C above the pre-industrial level. This two-degree target was set as a notional threshold above which 'dangerous' climate change is likely to occur. An aspirational lower target of 1.5°C was also set, as many countries, particularly low-lying island states, successfully argued that 2°C will lead to sea level rise that will make their islands uninhabitable.

Unfortunately, global emissions continued to increase from 2015 to 2019, and the UN predicts that if we continue to follow this emissions trajectory then this could lead to an increase of over 3°C by 2100. Even worse, these predictions are inherently cautious as the Intergovernmental Panel on Climate Change draws on the weight of scientific evidence in their forecasts. This means that they do not incorporate the latest (uncorroborated) evidence and research on sea level rise, tipping points and changes in cloud cover – some of which are quite worrying.

One degree, two degrees, three degrees, four ...... What does it matter? Well, the world was around 5°C cooler during the peak of the last Ice Age. An ice sheet expanded to cover London, and ice buried Scotland to a depth of 1,000 metres. Our climate is clearly susceptible to major climate changes initiated from minor natural changes in our orbit around the

Sun. The logic follows that our climate is also susceptible to human impact.

Two degrees is not a magical panacea; a future to look forward to. Taking sea level rise alone, it will result in ice sheets such as Greenland and West Antarctica gradually melting then disintegrating over hundreds of years resulting in a slow but inexorable increase in global sea level. Huge glaciers, such as the remote Thwaites glacier in West Antarctica are particularly vulnerable as their ice lies on top of sea water rather than on land. This means that warmish water can readily melt the ice from below as well as warmer temperatures melting from above. Given this simple fact, since when did it seem sensible to limit warming to two degrees? Future generations forced out of central London, New York, Shanghai and Jakarta will not be pleased by our generation allowing the temperature to increase by two degrees.

The negative effects of climate change do not increase steadily with each increase of one degree. Instead they accelerate causing permanent damage. Imagine a group of young teenagers being forced to drink litre glasses of beer. One litre equates to a one-degree centigrade increase in temperature. The first half litre feels good. The group feels more sociable, life is good. After one litre they still feel good, but it is having a noticeable effect. It is not safe for any of them to drive as their judgement is impaired. After two litres they are not enjoying it. They feel dizzy and stumble into each other and other people. An argument and a fight break out. After three litres they cannot think straight, are sick and the alcohol causes some permanent brain damage. After four litres all the group is unconscious. An ambulance is called. The world has already warmed by $1.1°C$ since the start of the industrial revolution. Like the drinkers we will regret going above $1.5°C$.

Even worse, the predictions about increase in temperature refer to an average across the globe. Temperatures over land will increase more than those over oceans and temperatures in the Arctic will increase even further with devastating impacts on the permafrost and Arctic ecosystem.

****

To be able to make good climate choices we need to understand carbon. Scientists talk about carbon dioxide 'equivalent' which includes all human-made gases and their contribution to global warming. This is a complicated calculation that takes account of the vastly different potencies (or strengths) of each gas and estimated time they might remain in the atmosphere. Carbon dioxide is the main component, followed by methane and nitrous oxide. Although the volume of methane and nitrous oxide emitted by human activity is less than that of carbon dioxide, they are more potent at warming our atmosphere and are therefore significant. Over the commonly cited 100-year period, methane is 28 times and nitrous oxide 300 times worse than carbon dioxide. However, unlike carbon dioxide, methane breaks down in the atmosphere. Over the next critical two decades for climate change its impact is 84 times worse. Burning methane (natural gas) does prevent the methane from reaching the atmosphere, but this still emits carbon dioxide.

What does a tonne of carbon dioxide look like? For a start it is invisible, but like all gases it is quite light, although it is slightly denser than air. It weighs just 2kg per cubic meter compared with water which weighs 1000kg. So, a tonne of concentrated carbon dioxide would fill a cube with sides 8 metres long. This equates to a medium sized hot air balloon.

What do you need to do to emit one tonne of carbon dioxide? Well, this will depend on what is burnt to supply your energy or how your food is grown. More of this later, but to help you to get a feel for it, one tonne approximates to driving 6,000km in a diesel car, one passenger on a return flight from London to

New York, or the average household annual consumption of 4,000 units of electricity in the EU. In countries powered by renewable electricity such as Iceland you could consume more units of electricity for one tonne, but less in those that burn coal like Australia.

To compare international footprints, for carbon dioxide only, the average citizen of the USA and Australia emits around 16 tonnes per year; in Kuwait it is 24 tonnes; China 8 tonnes; the UK 6 tonnes; India 2 tonnes and Malawi 0.1 tonnes. The global average is 5 tonnes. So, an average person brought up in American emits 160 times that of a person who happens to live in Malawi. There is a huge global inequity here as countries such as Malawi have not caused climate change but given their poor infrastructure and susceptibility to drought will be vulnerable to its effects. Put another way, Europe and North America account for 16% of the world's population but emit 46% of global emissions.

The USA has high emissions per person because of its affluence and consumerism, low population density, large houses, and urban spread. Away from city centres, travel by car is ubiquitous, often in gas guzzling vehicles encouraged by low fuel prices and fuel efficiency standards for vehicles which are not as tight as those in Europe. Due to the vast distances and poor rail infrastructure, many regularly take domestic flights. Meanwhile Australia is highly dependent on coal. Whilst emissions in the USA and Europe are falling, those in China have risen rapidly, and in India they are also increasing, albeit from a lower start point.

However, the above figures only account for carbon dioxide and the emissions that arise within a country such as producing electricity, burning gas, growing food and producing goods consumed in the country. As these are relatively easy to measure and allocate, these so-called production emissions are what countries tend to discuss at global climate change negotiations - our UK and Scottish targets are based on our production emissions. However, a more realistic and useful measure is the total emissions arising from our consumption of goods and services. This is our production emissions plus those associated with goods that we import, less emissions embedded within any goods exported. This more accurately reflects the impact of affluent lifestyles in high income countries that rely so heavily on importing materials and goods, often from low income countries. But measuring consumption emissions is inherently more complex as it requires an understanding of where goods and commodities are sourced from and the carbon intensity of their production. Given the reliance on imported manufactured goods, consumption emissions doubles the total carbon emitted by countries like the UK making the global inequity in carbon emissions even wider. In 2017, the UK Government estimated the consumption footprint was around 12 tonnes per person and even this excludes the emissions from extracting oil and gas from the North Sea.

Politicians in high income countries take advantage of accounting tricks to claim how their countries have cut emissions dramatically since 1990. These claims are invariably based on production emissions and often exclude international shipping and aviation. In reality, the cuts are not as high as many high income countries have increased their proportion of outsourced emissions as manufacturing and food production move overseas.

On the internet you can read all sorts of figures about average carbon footprints of individuals in different countries. These use different starting years, different categories to divide emissions into, some do not include all greenhouse gases, and some include or exclude international aviation and positive and negative emissions from changes in land-use. The detail is not too important, but the chart below provides a flavour of the carbon footprint of an average person living in the UK using an amalgam of different sources and a dose of common sense. It divides total emissions into the main categories of goods and services that we consume as individuals. The figures include all greenhouse gases and represent the 12 tonnes average emissions attributable to an individual living in the UK in 2020.

Figure 1: UK Consumption Emissions (source: author from various amalgamated sources)

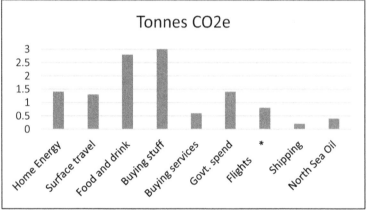

* uplift of x1.9 applied to the carbon emissions from flying to take account of other warming impacts, such as contrails forming clouds at altitude (as advised by BEIS, UK Government)

In the UK, the most significant emissions arise from us buying 'stuff' (houses, cars, clothes etc), eating and drinking, our travel, heating our homes and government expenditure on schools and hospitals. Emissions from our use of electricity have fallen dramatically because nuclear and renewable sources have replaced coal, so home carbon emissions are now mainly from heating our homes. Flying for leisure purposes accounts for around 6% and has been increasing, but this is an average. Many people never take a flight, for others, flights will be their single largest contribution to climate change. It is important to grasp that just as there is a vast difference in emissions between countries; within countries there are people who live high carbon lifestyles and those who do not. Examples of the former are those who fly for work or leisure regularly, drive a lot, own a second home, eat a lot of meat and enjoy regular leisure activities that involve travel or burning fossil fuel. In the UK, a relatively low carbon lifestyle might be someone who never takes a flight, lives in a small shared flat, walks or cycles to work and is a vegan. The top 10% of affluent consumers account for half of all carbon emissions. Most live in high income countries, but a growing number are from countries such as China, India and the Middle East.

It is also possible to calculate the carbon impact, or carbon footprint, of individual goods and services, although these will be rough estimates. It is particularly difficult to accurately measure the emissions from agriculture and land-use change.

Consider the difficulty in assessing the impact of a cup of coffee bought in a cafe. Emissions arise from growing coffee beans - fertiliser and harvesting; to process and sell the product - store, process, package, transport and retail; then to make the cup of coffee in the cafe - boil the water and to manufacture the single use plastic cup (and its lid, holder and stirrer), or to wash a reusable cup. Even the water from the tap has been collected, filtered, transported and had chemicals added to purify it. Then there are emissions from sugar, milk and the ubiquitous napkin(s). Emissions for milk depend on how the cattle are raised but include growing feedstuff, methane from cows burping, nitrous oxide from manure, oil for tractors, electricity for milking and refrigeration, transport, processing, packaging and retail. The result is that emissions from one cup of coffee can vary between 20g and 340g. At the lower end of this wide spectrum is a cup of black coffee with no wasted boiled water, at the upper end is a large and milky latte. Milk is a high carbon product, primarily because of the methane emitted by dairy cows.

Similar calculations can be made for all sorts of choices that we make. These can get complicated, but a modern fast electric hand dryer is better than using paper towels. Lightweight plastic bags have a lower carbon footprint than cotton bags due to the fertiliser used to grow cotton. We need to reuse bags to get the most benefit out of them.

Every tonne of carbon dioxide will contribute to warmer temperatures, ocean acidification and climate change. So, the logic is that the emissions we create will lead to someone, somewhere being flooded, being forced to emigrate or lose their livelihood; for a square metre of Arctic ice to disappear for ever; and for wildlife to be adversely affected eventually leading to the extinction of species. It is sobering to think that our actions indirectly cause so many difficulties to people and wildlife elsewhere. In most other circumstances we would be embarrassed, if not morally opposed, to knowingly cause harm to other people or to wildlife.

Meanwhile, climate scientists at Oxford University and the Potsdam Institute calculated that the maximum carbon that humans can 'safely' emit is approximately 1,000 billion, or one trillion tonnes. This would give a 50% chance of the Earth's climate not rising above the so-called dangerous threshold of two degrees centigrade. To reduce this risk further we should cap our emissions at 750 billion tonnes. So far, human activity has raised global temperatures from the pre-industrial average by 1.1°C and adverse impacts are already being felt. This warming would be even higher if the dust and air pollution from countries like China were not acting to temporarily reduce regional temperatures by reflecting solar radiation back to space.

One trillion (1,000,000,000,000) is a big number, but not big enough. Humans have already emitted over 600 billion tonnes, but our emissions only really became significant from the 1950s onwards. In 2018 alone humans emitted more than

40 billion tonnes of carbon dioxide. Historically the oceans, soils and forests have absorbed around half of our emissions, but we are pushing these carbon stores to their limit. Some scientists have even calculated that we have ten years left before we breach one trillion. The carbon budget for the planet is of course controversial, but whatever the exact figure, we are rapidly reaching an uncomfortable threshold, a point of no return. In fact, there is no magical date by which we need to cut emissions - like a smoker, every extra cigarette may reduce our health and life expectancy. Put simply, we need to cut emissions soon and fast, as every fraction of a degree centigrade will make a difference.

Another way to look at this is that we cannot burn all the coal, oil and gas reserves that we have already discovered. If we did, we would be well over the one trillion tonnes limit, perhaps two or three times over. We need to keep over half of the currently known economically recoverable reserves in the ground. This should be a wakeup call to investors, to employees in these industries and to all of us who continue to use fossil fuels. We must act soon and act fast.

Chapter 4:
The Impact of Humans

In simple terms the impact of humans on our planet can be described using the formula:

Global impact = Population x Consumption x Carbon intensity x Land used

If you were to consider the emissions from a country, then you could also add imports and extract exports. You could also deduct carbon sequestration, the removal of carbon dioxide and its safe burial underground or in soils or vegetation. But let us keep it simple for the moment.

****

Global population soared during the 20th century. When I was born in 1966 there were 3.4 billion people. When I studied geography at university there were 5 billion. There were 7.7 billion humans in 2019. It is difficult to appreciate the scale of this population growth, but there is a digital counter in Dynamic Earth - an educational tourist attraction in Edinburgh. You can watch the counter constantly going up. Our population increases by 220,000 people *every* day. That is every day last year, this year, and next. A sad but interesting fact is that the tsunami in the Indian Ocean on Boxing Day 2004 is the only day in recent human history when the global population fell.

As explained in more detail in a later chapter on agriculture, the world produces more than double the amount of food that we need to consume to keep healthy. So, the lack of food across the planet does not act as the brake on population growth that you might expect.

The United Nations projections vary greatly but they estimate a population of between 9.4 and 12.7 billion by 2100. Some think that population will fall thereafter but this is uncertain. Much of the growth will be in sub-Saharan Africa, forecast to double from 1 to 2 billion by 2100. Because of its susceptibility to drought, this is a region that climate change will heavily impact.

India is set to surpass China as the world's most populous country by the end of this decade, with 1.5 billion people. India's growth rate has levelled off due to rising affluence, growing urbanization, advances in women's education and family planning. However, even as fertility rates fall to the replacement level of 2.1 babies per woman, the total population continues to rise in low income countries because a high proportion of the population is young, either at childbearing age or soon to be. Just like it is difficult to stop an oil tanker, it is difficult to reverse this in-built population momentum. Unlike changes in consumption patterns which could take place quickly, any population control policies will take decades to have a significant impact.

In the long-term it appears that economic development is the best contraceptive. Fertility falls as childhood mortality falls and as countries develop. Women's rights, literacy, education and job prospects are important and of course the availability of contraceptives helps. Certain religious beliefs also act against population control.

So, population growth is a huge burden that strains our natural resources and adds to our emissions. Population growth impacts heavily on the local environment, particularly in regions short of water. We should do all we can to humanly reduce population growth then to reduce the population in a controlled manner. But I would argue that population growth is not the main problem. Bear in mind that the average person in Europe causes the same emissions as over 100 people in Malawi. So, our lifestyles and consumption are important, and it is high income countries that have caused climate change. Blaming the issue on population growth in low income countries is a diversion from the real issue of over consumption.

****

Consumption is the elephant in the room – a large and visible problem, but usually ignored. Producing physical goods (or 'stuff') and rampant consumerism is a recent phenomenon, originating in western countries, followed more recently by China. Politicians in low income countries are keen for their populations to catch up.

All forms of consumption have soared since the end of the Second World War. The rate of growth has accelerated. One estimate suggests that humans consumed 27 billion tonnes of materials in 1970. This has now increased four-fold to 110 billion and is projected to increase to 170 billion by 2050. In the three years of rapid urbanised growth to 2014, China used as much cement and concrete as the USA did in the preceding 100 years - Bill Gates, the founder of Microsoft, called this "the most staggering statistic". Around two million tonnes of plastic were produced in 1950, 450 million in 2015. The plastic industry, like every other industry and business sector, has the aspiration to grow further. Multi-national companies are still building petro-chemical plants that will lock us in to big increases in the volume of plastics manufactured. Less than 10% of all the plastic ever produced has been recycled. The rest is burnt, dumped in landfill, or too often, ends up as litter on land or discarded in the oceans.

Many people in high income countries live absurdly consumerist lifestyles. I was horrified when I counted the number of shirts in my wardrobe. A mix of 20 long and short-sleeved business shirts, 1 smart shirt to wear with a kilt, 6 warm long and 4 short-sleeved for outdoor use and hobbies, 5 long-sleeved shirts and 6 casual ones, 15 t-shirts, 6 lightweight summer holiday shirts, 2 old shirts used for painting or gardening, 4 long thermal and 5 short-sleeved thermal shirts for hill-walking and 6 lightweight sports shirts, including tennis, orienteering and running. 80 in total, and I do not consider myself as fashion-conscious, an excessive consumer or wasteful in comparison to many others. Of course, I am reluctant to throw wearable shirts out, and I received half of them as presents or as 'free' rewards for completing sports events.

Affluent people are drowning in 'stuff'. We fill our cupboards, our attics, our garages; some even pay to store stuff out with their homes. People buy disposable tents, fashion clothes they wear once, single-use plastic containers, they receive and throw out free plastic toys, they own two cars, and pay for non-essential cosmetic surgery. Children and students have en suite bedrooms. Goods are over packaged, for example, you can buy Easter eggs with five layers of wrapping. The manufacturer wraps each sweet in plastic wrapping, within a clear polythene bag, inside a chocolate egg wrapped in aluminium foil, inside a plastic container then placed by the retailer inside a carrier bag. People waste energy by lighting their homes when they are not in and even light their gardens at Halloween and Christmas. We are over consuming with no end in sight - you can now reserve a space on a tourist flight into space.

Many people also over consume food. If we eat too much, we become obese and less healthy. Food production takes up land, and requires fertiliser, pesticide and insecticide, diesel for tractors, energy for food processing and transport to shops.

Many of the goods we buy such as tobacco and cotton are made from agricultural products with the same requirements as for producing food. Paper and toilet tissue are made from trees. Natural forests have been cut down and replaced by fast-growing monocultures with devastating impacts on wildlife. America accounts for 4% of the world's population but consumes 20% of all toilet paper and tissue. Most is from virgin pulp, sourced from clear-cutting large areas of coniferous forest in Canada. As other countries become wealthier so does their demand for goods such as toilet paper and tissue. The demand for such products seems to be insatiable.

Other goods are made from plastics, composite materials or metals such as steel. These are made from oil and gas, metals and minerals mined from the ground and need energy to manufacture powered by coal, uranium, gas or oil. Over half of all oil is used for transport, 10% to heat buildings, 10% by industry; with over 10% used by the chemical industry to make products such as fertiliser, synthetic clothes, lubricants, bitumen and tyres, and a further 5% to make plastics.

Constructing our infrastructure of roads and buildings requires materials such as glass, concrete, steel and asphalt. Producing all of these is energy intensive and environmentally damaging. Concrete is made from cement and sand. Cement requires energy to heat the main ingredient limestone, and a chemical reaction in the process emits more carbon dioxide. Sand and aggregates are extracted from surface quarries or from dredging rivers.

Mining is energy intensive, disruptive and destructive to the local environment. It often causes pollution nearby or downstream and is usually visually unattractive. Building new access roads to mines opens wilderness and land owned by indigenous groups to further development and encourages further encroachment by agriculture. Like an iceberg, we only see the tip of mining activity. Most impacts are hidden out of sight in remote regions, often in low income countries with cheaper labour and often with a lax attitude to environmental protection, workers' safety and employment rights.

The Global Footprint Network estimates that if everyone on our planet consumed the same as the average Briton, we would need three planet Earth's to provide enough food, water and minerals to maintain our lifestyles. Of course, not everyone consumes as much as us (although some consume more), but the world population is growing, and most low income countries aim to catch up to our standard of living. The calculations behind this are complex and somewhat controversial, but it is difficult to believe that a growing global population can all consume the volume of physical goods and services that we currently think we need and believe we enjoy. Efficiencies and new technology will of course help but will not be enough.

So, the more food, goods or energy intensive services we consume the more impact we have on our climate and on wildlife. We need to find a way to satisfy our needs without consuming as much. Put simply we need to live with only one-third of our current impact, which will require a reduction in the volume of physical stuff that we consume.

****

Carbon intensity is the third aspect to the global impact formula. It is defined as the emissions per unit of energy input. This depends on the source of fuel and how efficiently that fuel is used. Electricity generated from coal emits around twice as much carbon dioxide per unit of energy than electricity generated from gas. Electricity generated from renewable sources is much better, although not perfect, as there are still emissions from mining and producing steel, rare earth minerals and silicon for wind turbines and solar panels.

Energy efficiency helps to reduce the carbon intensity of energy and benefits society as it enables the production of more goods or services from each unit of energy. Energy efficiency is further considered in a later chapter on resource efficiency.

****

Land used is the fourth aspect. All consumption of goods and services involves an impact on land-use either directly or indirectly - land that might otherwise be available to wildlife. Around one third of the land area is barren ice, desert or mountains; one third forest and one third is agriculture, mostly pasture. Humans have significantly altered over half of the land area of planet Earth, including many forests.

We grow most of our staple crops on land that was once forest. Whilst Europe cut down its trees hundreds of years ago, deforestation continues today in low income countries. Palm oil plantations are planted on land that was tropical rainforest and farmers continue to burn the Amazon rainforest to convert it to farmland often to grow crops to export to high income countries.

Our civilisation depends on the fertile soils that cover about one-tenth of Earth's land area. Soil is lost or degraded by erosion, water logging, compaction and salinisation. Like climate change, soil degradation is a long-term, mostly invisible problem that governments pay little attention to. Soil is a fragile skin, typically 15cm deep, but it comprises a complex biological ecosystem. Poor-quality, or overused soil holds less water, organic matter and nutrients leading to reduced crop productivity. The Great Plains of the USA originally contained rich, dark organic matter from the decomposition of prairie grass. Then ploughing, followed by a severe drought in the 1930's, loosened the soil enabling the wind to pick it up and create enormous dust storms. Today, the starkest problem is desertification in north-west China with drifting sand causing people to abandon farms and villages. The Sahel region of Africa is also particularly fragile from a combination of its harsh climate, population growth and increase in livestock numbers.

Deep ploughing can release carbon from the soil, whilst heavy machinery can compact and damage soil. Chemical fertiliser and pesticides kill organic matter that binds soil particles. The risk of erosion by wind and water is higher on steep slopes, bare fields and in drought prone areas. Over irrigation can result in salt rising through the soil making it infertile. Coastal salinisation is another serious problem that is difficult to reverse. Rising sea level combined with the over-extraction of freshwater from underground aquifers can enable saline water to percolate inland causing distress to most plants and crops.

In general, staple crops and vegetables grow efficiently and require far less land than products from livestock. Cattle require large areas to graze or to grow crops to feed them if they are kept in intensive enclosures. Food calories from animal protein require more land than those from plants because mammals burn calories to move around and to stay warm. Eating insects would be less land intensive than livestock.

Organic food has the significant environmental benefit of not using synthetic fertiliser, pesticides or insecticides, but it normally produces smaller yields per hectare than non-organic crops. On a planet wide scale, it is therefore possible that organic crops may cause more carbon impact although they undoubtedly have many environmental advantages. Similarly, there is an argument to support genetically modified crops if they produce higher yields with fewer artificial pesticides and insecticides. There are other solutions such as crop rotation and growing more legumes which fix nitrogen from the air and therefore need little nitrogen-based fertiliser. These are complicated decisions that are explored in later chapters on our diets and agriculture.

All goods are manufactured from materials which have been grown, quarried or mined, with consequent impacts on land-use. The UK once had thousands of quarries and mines, but most have closed, and we now import most raw materials. The problem is hidden. Invisible. But would we be happy to know that our overconsumption causes deforestation in the Amazon and displaces wildlife and indigenous people?

Finally, we should consider imports. We need to take care to avoid actions such as high taxes that may force manufacturing plants in the UK to close, as these raw materials are then likely to be imported by ship from low income countries where health and safety, labour laws, land-use and environmental regulations may not be as stringent.

So, the impact humans have on the environment is caused by our consumption of goods and services, and the energy, water and associated land-use needed to produce these. Population growth exacerbates the problem, but over the next few decades it will be easier to reduce consumption and to improve the efficiency of producing goods than it will be to reduce our population.

Chapter 5:
It's more than Carbon –
Wildlife too

The environmental crises we face are not just about our emissions and climate change. Humankind is also destroying nature and wildlife across our planet. If we continue to increase our use of natural resources, we will trigger the world's sixth mass extinction. We depend on nature for many of our resources and our life support systems, but more than that we have a moral duty to protect nature and wildlife. Earth is the only known planet with life on it in our galaxy. Humans have become the stewards of planet Earth. It should be unforgivable to destroy wildlife and cause species to become extinct. Future generations will not forgive us, just as we do not forgive the human settlers who decimated native wildlife on every new island they colonised. The flightless dodo is a small symbol of this – it became extinct through hunting and the introduction of non-native species such as rats, cats and pigs.

There is a complex interplay between land-use, climate change and wildlife. Each affects the other. For example, chopping down forests and overexploiting soil from intensive agriculture both release carbon. Grasslands, coral reefs, mangroves and peatlands store carbon. Human activity can release carbon by degrading or destroying these habitats; conversely, we can remove carbon from the atmosphere by protecting and restoring these habitats, which will also benefit wildlife. Even hunting whales has an impact. Whale faeces, rich in iron and nitrogen, provides food to plankton which absorb carbon dioxide removing it from the atmosphere.

Nature is integral to this book. Although we might be able to change our lifestyles and deploy technology to solve climate change, if all we achieve is to find low carbon ways to over-exploit natural resources we will live in a depleted, diminished planet with wildlife and nature decimated. And we will lose many of the essential services that nature provides us for free; clean water, ingredients for new pharmaceuticals and pollination of fruit and crops.

The impact of humans on wildlife is shocking and astonishing. We have reduced the total global animal biomass (its weight) to one-sixth of what it was; whilst the biomass of all wildlife, including plants, has halved. Human beings now account for a third of all mammal biomass; our livestock account for two-thirds, leaving wild animals at only 4%. Put another way, livestock outweigh wild mammals by 15 to 1. Poultry accounts for three times the biomass of wild birds. Only 15% of global wetlands remain in good condition and humans have removed most natural forests or replaced them with monoculture plantations devoid of species diversity. There are three trillion trees today, around 400 for every person, yet humans inherited a world with twice that number. There are 20,000 native bee species that pollinate 70% of all plants. Many have evolved specialisms with individual plant species so if one becomes locally extinct the other will too. But native bees are in decline due to intensive agriculture - their pollination services are often replaced by farmers with large colonies of domesticated honey bees – native to south-east Asia – reducing the biodiversity which is essential to maintain ecosystems.

The main cause of this catastrophic decline in wildlife is the fragmentation and loss of habitat. Converting land to agriculture is the main driver of this. Deforestation is a key part, whether it is removing trees for agriculture or replacing a wildlife rich forest with a monoculture forest plantation. Urbanisation is an even more thorough process at destroying wildlife as it covers much of the ground with tarmac and concrete.

The drivers of land-use change vary from country to country and over time. The Second World War disrupted food imports from the British Empire that the UK relied on. The population was traumatised by the fear of food rationing. It became the patriotic duty of farmers to maximise food production. They ploughed pasture to grow crops, drained the land, removed trees and hedgerows, filled in ponds, straightened rivers, and ploughed wildflower meadows. These deliberate actions accelerated after the war promoted by subsidies, firstly from the UK Government then from the EU, paid to maximise production. This encouraged more use of fertiliser, pesticide and insecticide. The largest payments went to the largest and wealthiest landowners. We were left with a neat and tidy landscape with large fields but little scrub, wildflower meadows, ponds, hedges or trees. Drainage accelerated the run-off of water from fields leading to a higher risk of floods downstream.

Where humans over-stock herbivores this leads to a permanent loss of tree cover and ecological devastation. Sheep in the UK and Iceland, goats in southern Europe and the Middle East, rabbits in Australia. But much of the ecological damage inflicted by human activity is invisible to the layperson. Only a historian or ecologist truly understands what we have lost. Tourists enjoy these bare 'wilderness' landscapes not realising that they are human-made and the result of an ecological catastrophe.

In the Amazon, the conversion of land to pasture and to grow soya to feed cattle and chicken drives deforestation of the tropical forest. Mining is important in some areas and dams with vast new reservoirs destroy the forest too. Decaying vegetation in the new reservoirs release methane. In Asia, the main driver is the conversion of tropical forest to monoculture palm oil plantations. This can also displace subsistence farmers onto more marginal land. In Madagascar the spread of subsistence farming and logging are threatening species of lemur with extinction. Affluent consumers drive these processes by buying processed foods and cosmetics. However, in a few areas of the world, this process of deforestation has reversed with young people abandoning marginal agricultural lands as they migrate to cities. This can give wildlife a chance for a comeback unless a species has already been driven to local extinction.

Wildlife is also in decline due to hunting, poaching, pollution, and diseases spread by humans and livestock. Some non-native species, whether introduced by accident or deliberately by humans can spread and dominate a landscape if there are no natural predators. This is an insidious, long-term issue that is expensive, or in many cases impossible to control. No matter where you live in the world, non-native species will already be affecting the natural world around you. Like an invader's army, navy and air-force; this stealthy invasion marches across our land, soil, rivers, seas and in the air. Their global spread is accelerating due to the increase in international trade, travel by humans and from organisms hitch-hiking a lift on transport such as ships. Whilst not all non-native species are invasive, a small number are, and these can devastate native wildlife and impact our economy.

If that was not enough, we are now introducing a new mechanism to decimate wildlife. Climate change is a 'threat multiplier'. It adds an additional pressure to what might already be a marginal situation for wildlife.

****

As an example that I am familiar with, the Scottish Highlands draw tourists to see the scenery and to admire what they see as a natural wilderness. But the iconic heather moorland that dominates the Eastern Highlands is human made. In Scotland, large landowners manage 12% of the total land area for grouse shooting and 14% for commercial deer stalking. There are six million sheep in the lowlands and uplands. Deer and sheep are both ruthless grazers which prevent any tree saplings from growing. Fewer than 500 people own over half the land in Scotland, mostly in large estates in the uplands. Most of these sporting estates are set amidst spectacular landscapes but are neglected from an ecological perspective. Land has been drained, drying out the peat and contributing to erosion which harms salmon and can cause flash flooding. Landowners burn and control the land to provide a human made, heather dominated landscape to suit grouse. Centuries of persecution, now mostly illegal, have reduced the number of mammal predators and birds of prey. Gamekeepers even shoot mountain hares as they carry ticks which may harm grouse. The soil and vegetation have been exhausted and deer living in open moorland grow to be smaller than those living in natural woodland.

The natural Highland landscape would be a patchwork of forest, dominated by hazel, ash and alder near the west coast and silver birch and pine in the drier east but intermixed with other species. Trees would gradually dwindle in height and grow further apart high on the mountain sides but even in the valleys the forest would not be continuous. Herbivores such as horses, wild cattle, deer and boar all graze differently creating bare patches and opportunities for wildlife in different habitats. Ancient hollowed out trees would provide habitat for birds, insects and bats. Rivers would meander through the glens with wetlands created by beaver dams whilst predators such as wolves and lynx would prevent overgrazing by keeping herbivores restless and constantly on the move.

Virtually none of this forest remains with only 4% of the land covered in semi-native trees. Conservationists argue over the extent which climate change, natural succession, overgrazing and other impacts of humans has caused deforestation and species loss. But in recent centuries landowners forced the indigenous people off their ancestral lands during the Highland Clearances to make way for sheep and deer. Grazing prevents tree saplings from growing to this day. The only truly natural land is the high mountain tops, some peatland and fragments of sea cliffs, dunes and salt marsh. Over centuries, the wolf, lynx, bear, elk, wild boar, beaver, auroch, crane and capercaillie were all hunted or driven to extinction. Victorian Highland estates ruthlessly eliminated natural predators to maximise the number of grouse or deer on their land. Today the main predator is domestic and feral cats. Otters declined due to hunting, poor water quality and habitat destruction. The salmon population is in sharp decline for several reasons including climate change affecting its migratory patterns and access to food. If rivers warm up too much many fish will not be able to breed.

Dense plantations of Sitka spruce from America shade out any undergrowth, whilst grey squirrels, also native to America, have invaded. Rhododendrons have spread from the Victorian gardens where they were planted as an ornamental plant. They are rapidly spreading downwind and along road verges with the seeds carried by turbulence from passing cars. Rhododendrons are particularly effective as a non-native invasive plant as their dense shade smothers other plants, their leaves are poisonous, and they exude toxins which suppress germination by rival species. Along riverbanks, Himalayan balsam, with its pretty purple flowers, is spreading out of control as is giant hogweed despite repeated spraying by under resourced councils and volunteers. Its dense thickets out compete other plants and in winter it dies back leaving bare soil prone to erosion. But the real danger is that if you touch its sap it makes your skin hypersensitive to sunlight.

45% of Europe's seabirds live in Scotland. Although most breed within areas of special protection status, many species are still in decline. Arctic Skuas are at the southerly limit of their climatic range. Puffins rely on eating sand eels which have been fished on an industrial scale to produce fish meal for salmon farms, whilst the sand eels are susceptible to changes in sea surface temperature. Non-native rats, cats, stoats and even hedgehogs eat bird eggs and effectively control the range of bird species restricting breeding success. Oil pollution, ingesting plastic and even wind turbines also impact on seabirds, but these are minor compared with climate change and the associated shifts in the distribution of prey.

This might seem bleak. It is, but the big picture is not good in Scotland or globally. But there are some amazing pockets of good practice which demonstrate what can be done. A handful of species have been reintroduced including the white sea-eagle and most recently the beaver, despite opposition from many farmers and landowners. Returning land to nature, or rewilding, is gaining a slow but unstoppable momentum. The most ambitious project in Scotland is the 600km^2 Cairngorms Connect project where private and public land managers have a 200-year vision to enhance habitats, species and ecological processes within the Cairngorms National Park. The transformational results can be fast and exciting. If herbivores are controlled, then much of the regeneration can take place without further human intervention. Within a few years, insects and birds return to areas where they have been absent for centuries.

A final thought on wildlife. Wildlife needs space and access to water. And lots of space. A thriving ecosystem needs a range of wildlife including large herbivores and predators. But human activity can easily disturb predators. Globally, big cats, such as Siberian tigers need a wide range to hunt and access to neighbouring ranges to prevent the risk of genetic inbreeding. Fragmentation of their hunting forests creates 'islands' of habitat which are hopelessly unsuited to long-term conservation. The book, 'The Great Soul of Siberia' by Sooyong Park, gives an evocative description of the dedicated and heart-breaking attempt to save some of the last surviving Siberian tigers in their ever-decreasing wilderness.

Section Two:
Ten Building Blocks to make better Carbon Choices

Introduction

The problems may sound overwhelming. But our aims can be clearly articulated - to phase out our use of fossil fuels, to be sensible about land-use and to restore the planet's ecosystems. Humans have successfully solved other environmental issues when we put our minds and resources to it. Many adverse impacts are part of a vicious spiral of decline; conversely changes can create a virtuous circle where one improvement leads to another. There are many examples of such success. Planting trees in sub-Saharan Africa stores carbon, protects the soil from heavy rainfall, provides shade and habitat for wildlife, creates a sustainable source of fuel for local people and may even prevent the spread of deserts. Electric buses are quieter and emit no air pollution which will make our cities more pleasant to live in. This will encourage cycling which helps to prevent obesity and improves physical and mental health.

We need to think widely about the impact of our decisions and actions and consider the knock-on consequences. Carbon footprinting helps us to understand our carbon emissions, whilst a wider environmental lifecycle analysis also considers impacts on water, land-use and people. Thinking about the whole system will help us to make better choices.

Perhaps in an ideal world business would only offer us 'green' choices, but in the meantime how can consumers hope to make sensible choices if manufacturers and retailers do not inform us of the environmental impact of their products?

The following ten chapters outline the building blocks that society needs to get right before we can successfully tackle climate change and wildlife loss:

- think long-term
- economics
- sensible regulations
- good design
- targeted innovation
- targeted investment
- education, jobs and training
- behaviour change
- involving communities
- resource efficiency

Think of these ten building blocks as the foundations to help us build a low carbon economy that works in harmony with nature. Some individual businesses and consumers currently try to make good decisions. But this minority finds it difficult as there are many obstacles. We need to clear these. Governments can set the policy direction and introduce sensible regulations, businesses can respond by providing innovative low carbon products and services, and consumers will have the knowledge to make better carbon choices.

Chapter 6:
Think Long-term

We cannot tackle climate change and wildlife loss without government, business and consumers all taking a long-term approach to decision making. For example, investing in insulation in houses increases our comfort and reduces fuel bills, but it may have a ten-year economic payback. If we took a short-term approach we would not make such investments resulting in us continuing to emit carbon with its adverse impact on current and future generations. A short-term approach often results in sub-optimal decisions, such as paying the lowest possible price to resurface a road only for potholes to quickly appear, requiring further expensive and disruptive road works in the future.

"Think and plan for seven generations ahead" is a saying attributed to Native Americans. This is quite a long-time, say between 140 and 210 years. It might be more helpful to think of your grandchildren's generation. Will you be able to justify your current lifestyle to them when they ask you what you did to prevent catastrophic climate change?

There are examples of society taking a long-term approach. In the UK, the Forestry Commission was set up in 1919 after the First World War with a remit to ensure a strategic supply of timber in the event of any future war. In Medieval times, the Forest of Dean was planted to provide oak to build warships. Although times have changed these were far sighted strategic decisions. A great example of a vision backed up by a long-term approach is the campaign, initiated by Rotary International in 1985, to eradicate polio. Its incidence has decreased by over 99% and in 2020 it was announced that the disease has been eliminated from Africa.

Long term planning can give direction and gives time for business and consumers to prepare. To tackle smog, California set the most stringent air pollution standards in the world in the 1970s which led to the development of catalytic converters. A current example is the UK Government's ban on new fossil fuel cars by 2040 (then changed to 2035). Although this is far away (too far) it gives car manufacturers time to adapt their fleet, and time for local authorities and power companies to set up the necessary charging infrastructure.

Similarly, the EU set a long-term target for Europe's car manufacturers. Average emissions of carbon dioxide from their fleets sold in 2021 must be less than 95 grams per km, down from an average of 160 grams in 2006. This legislation has pressurised car makers to invest in the innovation and the capital investment needed to bring new electric car models to the market. This pressure did not come from consumers.

However, short electoral cycles dominate politics in democratic societies which perpetuates short-term decision making. Even worse, most government ministers do not even survive one political term, certainly not in the same post. They are desperate to make their mark, so they implement incremental changes quickly, and avoid addressing complex long-term issues. In the UK, like other countries, issues such as financing care for the elderly, pension reform and making adequate preparations for a pandemic are never properly analysed and tackled.

The Climate Change Acts in the UK and Scotland are rare examples of legislation that received cross-party support for policies that extend to 2050 - well beyond our normal five-year electoral cycle. To overcome the recognised shortcomings of parliamentary democracy, the Welsh Government enacted a Wellbeing of Future Generations Act in 2015 to improve social, economic, environmental and cultural wellbeing. A commissioner was appointed to act as a guardian for future generations and to help achieve wellbeing. In practice, the commissioner's main role is to advise on the long-term effects and any knock-on impacts of decisions and policies made by public bodies. The long-term and strategic thinking behind this Act is absolutely in line with my thoughts around overcoming the existing barriers to tackle climate change and wildlife loss.

But, generally as a society and as individuals we are not good at long-term planning. Many are reluctant to save for their retirement even though the government and financial advisors recommend it. We still build houses on floodplains despite knowing that climate change forecasts are for more intense rainfall, and we have locked ourselves into a rising sea level for hundreds of years. The waste from nuclear power stations will be radioactive for thousands of years.

The spread of damaging invasive species is a classic example of the absence of a long-term approach. When the first seeds or invasive animals appear they could easily be eradicated. By the time anyone takes notice, or acts, their spread has accelerated. After several years, the cost to eradicate them spirals, and after a few decades it is unrealistic to do anything other than control them in certain places. Giant hogweed is an example where the council and volunteers spray weed killer, but the funding is short-term and when it runs out the hogweed recovers rapidly. The seeds can lie dormant in the soil for 20 years, so it requires dedication and a long-term approach to tackle.

****

We all choose to do things that emit carbon even although we now know that it will have a detrimental long-term effect. We could spend money to insulate our homes to the highest standard to cut our gas consumption. This might be cost effective over the long-term even using current low gas costs. But how many of us have done this? Other actions to reduce our emissions will be more costly, or at least are more costly at present. Electric cars are more expensive to buy than conventional cars, but their running costs are lower. Over the medium term the total cost of running an electric car might be similar to a conventional car, but people either do not have the finance or may not want to spend a lot of money to pay the high purchase price. On a more prosaic level, how many of us have bought a cheap umbrella only for it to break the first time we use it on a windy day? It would be better to buy a more expensive, better quality umbrella, built to last.

The average tenure of a chief executive of a company in the UK is less than five years. Like our politicians, they want to make their mark by enacting changes that they can make quickly. Even worse, shareholders and financial institutions demand short term returns on their investments and directors' pay is often linked to short term changes in the company share price. Consequently, companies are risk averse and adopt a short-term approach, for example they will only invest in energy efficiency if it has a two-year payback or less. Investments in on-site renewables such as solar panels may have a long payback of say ten years. Directors rarely make such decisions, even though it would be good for society and for the long-term sustainability of their company.

Unilever has been a rare exception to this. Their Sustainable Living Plan has long-term targets that align economic, social and environmental considerations. This long-term and strategic approach enabled its employees to be creative, leading to innovation and the successful launch of new products including soaps made entirely from natural products and fully biodegradable tea bags.

Many decisions for our society need a long-term approach. Investment in a nuclear power station can only be cost effective if it operates for decades. In Scotland, planning permission has been granted to build a new pump storage hydro-electric scheme, which would help to balance the electricity grid, but investors will not invest due to the long-term political uncertainty around future energy policy. District heat - hot water pipes under roads to heat homes - is capital intensive to install and may take decades to pay for itself. Denmark invested in district heat after the global oil crisis in the 1970s and benefits from this far sighted decision today. This investment in infrastructure provides flexibility to use different fuel sources depending on cost, supply and changing societal needs. Initially coal power stations heated the hot water then they changed to burn wood chips, to recover waste heat and to use solar panels.

Unfortunately, we continue to 'lock' ourselves into new infrastructure that does not support a low carbon economy. New dual carriageways, airport expansion and gas fired heat and power plants. We build new detached houses without the highest insulation standards or electric car charge facilities, connected to the gas network and built far from public transport.

If we are to tackle climate change and the loss of wildlife, it is fundamental that we think and plan for the long-term. Carbon will stay in the atmosphere for decades, sea level rise may be unstoppable, and the extinction of a species is permanent. A clear vision and long-term approach will lead to better decision making. There is no magic bullet. Government, business and individuals need a change of attitude - think of your legacy. Businesses are already setting long-term aspirational targets to reduce their environmental impact. Could you justify your current work, lifestyle and actions to your grandchildren?

Chapter 7:
Economics: Price is King

Capitalism has been incredibly successful at creating wealth and lifting many people out of poverty. Even nominally communist countries such as China and Vietnam now embrace capitalism. The price of goods is set by the law of supply and demand. The problem is that there is no monetary cost placed on carbon emissions or the destruction of nature and loss of wildlife. The 'free' market takes advantage. The cost is picked up by the taxpayer, or by nobody in which case we suffer the consequences such as poor air quality, water pollution and degraded habitats. What we need is 'sensible capitalism', starting with a price for the goods and services that we buy that reflects the environmental damage that they cause.

Sensible capitalism is getting the price right for everything of value:

Tax the bad, subsidise the good

Examples abound of where the price is not right. Goods and services that are too cheap include natural gas, plastic, petrol in many countries, meat and chewing gum. Flying is too cheap particularly compared to train travel. Landowners can harvest and sell trees, but this can lead to increased runoff and flood damage to society downstream that may outweigh any income to the landowner.

In all these cases the taxpayer or community bear the cost to remedy the damage. This damage could be the climate impact of the product, the invisible impact of micro-plastics or the cost to taxpayers to remove chewing gum from pavements using high pressure water jets. Economists call this damage a 'pollution externality' and talk about 'the tragedy of the commons.'

Pollution externalities are where humans cause environmental harm to others because there is no regulation or price incentive to stop them. Examples of this abound such as emitting carbon dioxide or allowing dirty wastewater to enter rivers. To illustrate the tragedy of the commons, imagine a small rural village. No individual farmer in the village has any incentive to avoid grazing their animals on common land, so they all overgraze until the vegetation and soil are exhausted. In the short-term all the farmers benefit, but in the long-term all in the community lose. Emissions of carbon dioxide, methane and nitrous oxide to the atmosphere are the ultimate global tragedy of the commons. Each individual country has an economic interest to burn all their fossil fuels, but this is at the expense of climate change that will cause global suffering. Across the world we place an insufficient price, or no price, on this atmospheric pollution. We pass the cost and damage to other people and to future generations.

The International Monetary Fund estimated that governments supplied $296 billion in direct subsidies to support the production or consumption of fossil fuels in 2017. High income countries also give over $100 billion of potentially environmentally harmful support to farmers. They also calculated that the full social cost of these subsidies is far higher; around $4.7 trillion with China providing the most. This is based on the difference between the price that consumers pay and the estimated full cost of these products taking account of all the environmental damage they cause including deaths from air pollution and the impact of greenhouse gases. In effect governments are using taxpayers' money to destroy our world.

Globally, governments introduce subsidies for a variety of reasons, often with a specific goal in mind. Subsidies might encourage companies to explore for new oil reserves to help make a country more self-sufficient in energy. Often governments introduce a direct price subsidy, or exemption from normal taxes, to help poorer consumers to access energy or to buy fertiliser. Once in place, subsidies are politically difficult to remove. Riots occurred in India when their government tried to reduce subsidies on cooking gas. In contrast, Indonesia did phase out consumer fossil fuel subsidies by doing so gradually over several years and providing compensation to its poorest citizens.

The International Energy Agency has argued that eliminating fossil fuel subsidies is the single most cost-effective way to reduce carbon emissions. In 2016 the wealthy G7 nations agreed to phase out such subsidies but have made little progress. This is inexcusable. Interestingly, the definition of subsidies is wide ranging including the failure to levy 'normal' taxes, for example the UK charges a reduced rate of VAT on domestic electricity and gas. Iran provides the largest energy subsidy per person in the world. Gasoline is cheap, which leads to high consumption, terrible air quality in cities and hinders the deployment of solar energy.

International aviation benefits from many subsidies. In most countries, there is no tax on aviation fuel or VAT on ticket sales. Duty-free concessions also help to subsidise aviation. Within the EU, the Emissions Trading Scheme covers aviation, but free allowances are handed to airlines, and flights travelling outside the EU are exempt. The international shipping industry is also heavily subsidised, avoiding taxes on fuel, VAT and corporation tax.

The biggest fossil fuel subsidies in the UK are the reduced 5% VAT rate for electricity and gas for consumers, followed by tax breaks for oil and gas operators in the North Sea. Also there is a Winter Fuel Payment made to all pensioners regardless of their wealth and an additional Cold Weather Payment made during very cold weather - at least it targets people when they are most in need. In Scotland, the Coal Authority went bankrupt leaving the taxpayer with the liability to clean up the environmental damage left behind by open-cast coal mining – a retrospective subsidy. Meanwhile farmers, fishermen, road refrigeration hauliers and plant on construction sites benefit from 'red diesel', a £2.4 billion reduction in normal taxes on diesel. Governments either deny that these are subsidies or justify them as being necessary to help those who are less wealthy. But such subsidies encourage us to consume more energy. In the long-term it would be better to spend this money to insulate our homes or on innovation to reduce the demand for fuel, for example to design new electric or fuel-efficient tractors and fishing boats.

Government policies change over time dependent on political and economic priorities. Subsidies can be used to achieve different objectives, which may or may not benefit the environment. As an example, after the Second World War, subsidies were provided in the UK to maximise food production. Farmers responded by draining bogs and wetlands. They tore out hedgerows to enable easier access to fields by large machinery. The Flow Country in the north of Scotland is the largest blanket bog wetland in Europe and is rich in wildlife. In a push to increase commercial forestry, but in the wrong location, tax breaks in the 1980s encouraged landowners to dig drains and plant non-native conifer trees. These trees were never commercially viable due to the remote location, poor soil and exposure to strong winds. The Peatlands Partnership is now paying to reverse this damage by cutting down the trees and blocking the drainage channels to restore the bog.

Good subsidies provide economic, social and environmental benefit. Ideally, they are short-term, designed to overcome a 'market-failure', and can then be abolished as the market responds. In the UK, government backed payments to consumers to support renewable electricity led to a dramatic uptake in rooftop solar panels. Within three years, the cost to deploy panels with an output of 2kWh fell from £10,000 to £5,000 for a system with double the output. Now, in good locations solar panels can be deployed without subsidy even in the cloudy UK. Although the renewable industry complains that subsidies were reduced and then abolished, the subsidy succeeded in what was intended – to kick start a new industry. But more care must be taken to ensure that they do not cause an inefficient boom and bust for installers. Now subsidies have moved on and are available to install electric charge points and to buy electric cars.

Subsidies can also help to direct business to be more sustainable. One example is innovation grants to develop new products such as more efficient batteries or stronger or lighter materials to construct wind turbine blades. Business can also be offered interest free loans to invest in energy efficiency measures to overcome their reluctance to spend scarce capital resources on projects with a long payback.

Taxes can influence behaviour and can be designed to take account of the adverse environmental impact of a product. Governments like 'sin taxes' such as those on cigarettes, alcohol, sugar and fuel as they are seen as an easy way to raise revenue and can be justified to help pay for health services. The UK's Landfill Tax is an example of a successful tax which led to a massive reduction in landfill in favour of recycling. Vehicle Excise Duty (car tax) has been based on a car's carbon emissions since 2001. This incentivised car manufacturers to design and sell more fuel-efficient cars and encourages consumers to choose such cars. However, the tax bands should be regularly tightened to take account of the ever-increasing fuel efficiency of new cars, otherwise the incentive to continually design more efficient cars will decline and the total amount of tax collected will fall. The Carrier Bag Charge in Scotland resulted in consumers using 650 million fewer plastic bags per year, a reduction of over 80%. There is a small benefit for the climate, but more substantial benefit in the form of less litter strewing the landscape. This is a great example of how to 'nudge' the behaviour of people towards better outcomes - and all because of a small, seemingly insignificant, charge of five pence per bag.

****

Our emissions of carbon dioxide, nitrous oxide and methane cause climate change. The logic, backed by most economists, would be to tax these emissions - a global carbon tax. Governments could apply the tax when fossil fuels are extracted from the ground or tax consumers when they buy goods or services. This would be an effective way to price emissions that would efficiently reduce our use of fossil fuels and steer us to choose lower carbon options. Unfortunately, based on current evidence, this is not going to happen. It seems to be politically impossible to reach an agreement across countries. Tax is something that governments like to control for themselves, and it is hard to see Saudi Arabia or the USA agreeing to tax their oil.

An alternative method of putting an effective price on carbon emissions is to place a limit or cap on total emissions. The European Emissions Trading Scheme (ETS) is an example, but other regions, including China have set up similar schemes. In effect, it is a form of rationing. Under the ETS the EU sets a cap on the number of tonnes of carbon dioxide that major industries can emit. It is set at a level where the demand to emit carbon is higher than the total allowed and therefore companies are forced to cut their emissions or buy carbon permits. This is an example of sensible capitalism, albeit administratively complicated. This scheme has had some success, but because of lobbying by industry, the cap is set too high resulting in a low carbon price that has had limited impact. Some sectors such as cement manufacture have even received 'free' allowances, arguing that if the cost of their energy is too high they will become uncompetitive against countries like China. The problem with the ETS is that it is not sufficiently robust, and the wider problem of all regional emission trading schemes is that they can be undermined by international trade.

Tax and refund are a third method put forward by economists to cut emissions. Here all carbon emissions are taxed. But instead of the government spending the tax income, they refund it direct to families in the form of 'cash' in their pockets, perhaps by paying an income tax refund. This protects low income families from the inevitable increase in the cost of basic services arising from the carbon tax.

Whatever tax regime governments implement, they have to act within limits loosely set by the expectations of society. When France tried to increase fuel taxes to help reduce its carbon emissions it led to weeks of protests and even riots in Paris. In Edinburgh, the public resoundingly defeated a referendum that proposed a congestion charge on drivers entering the city. The UK Government's 'fuel price escalator' steadily increased taxes to encourage more fuel-efficient cars and discourage car use. But after a few years the price had risen beyond a notional publicly acceptable limit leading to protests by self-employed lorry drivers who were disproportionately affected. In the aftermath, UK governments have not dared to raise this tax again. Similarly, the Landfill Tax encouraged recycling, but led to an increase in unsightly fly tipping. As taxes increase, certain sections of society campaign against them or find ways to circumvent paying them.

This suggests that it is important to consult with the public and try to gain consensus. The UK Government has set up a Citizens Assembly to gather the views of a random group of people to inform future government policy on climate change.

And, learning from the failure of Edinburgh City Council's referendum, Nottingham successfully introduced a potentially controversial Workplace Parking Levy. The council spends the income raised to improve cycle routes and to extend the tram network. The initial concern, that businesses in the city centre would leave, has not materialised, and in fact the city has flourished.

In the background, away from political intervention in taxes and subsidies, a quiet revolution is occurring. The price of energy generated from renewable sources has tumbled. And with new technology enabling more flexibility, the price of operating our electricity grid to cope with a high penetration of renewable energy continues to fall. This is beginning to, and will continue to, undermine fossil fuels. Developers are building the latest offshore wind farms with a far smaller subsidy than that required to build new nuclear power stations. And we will become less reliant for our energy from unstable regions such as the Middle East and Russia. In the future, the price of other new technologies, and solutions such as hydrogen made from renewable energy or carbon capture and storage may tumble too. This may open opportunities that we are currently blind to.

Abolishing subsidies and placing a sensible price on emitting carbon is an essential first step. It will change the economics of our energy choices, change mindsets and encourage innovation of novel solutions. But taxes are not enough. Businesses and consumers will not accept punitive taxes, especially if they are seen to be unfair compared with other countries. Taxes should be part of our armoury alongside sensible regulations, incentives to innovate and behavioural change.

Chapter 8:
Sensible Regulations

Regulations and bans on activities can drive improvement and stimulate innovation. They create a level playing field for businesses and steer, or force, consumers to better choices or outcomes. Regulation constrains almost everything we do. For example, seat belts, speed limits, drinking alcohol, lead in petrol and drinking water standards. Every time a new regulation is proposed there is a minority of forward-looking companies who welcome it, but also a trade body, representing vested interests, will often oppose, seek to delay or water it down. Car companies objected to the introduction of unleaded petrol, then to the introduction of catalytic converters and now object to the timetable for phasing out petrol and diesel cars. Consumers are confused, fuelled by alarmist newspaper headlines, and as often as not are initially against new regulations.

In practice, once enacted, the public and business quickly accept most regulations. Scotland banned smoking indoors in pubs in 2006. This ban was controversial at the time, yet few would now argue to revert to the past.

The European Union banned the use of incandescent light bulbs. Many newspapers were in uproar. These were the traditional light bulbs, or more accurately 'heat' bulbs as they wasted 90% of the energy as heat. Initially the replacement compact fluorescent bulbs were not good quality, and this led to some genuine consumer backlash. However, retailers soon introduced far more energy efficient Light Emitting Diodes (LEDs). These have slashed emissions from lighting our homes with cost savings to consumers too.

Similarly, the UK Government was frustrated that industry had developed more energy efficient condensing gas boilers, but few were sold. They therefore mandated all new gas boilers in our homes to meet higher levels of efficiency. The new condensing boilers operate at around 90% efficiency, replacing older boilers at 65 to 80%. This single regulation resulted in a measurable fall in UK gas demand and its associated carbon emissions even though it is taking years to fully take effect as households gradually replace their old boilers.

The EU introduced regulations to incentivise manufacturers to design and sell more energy efficient products. Rather than setting a minimum threshold, fridges and freezers were assessed and awarded a rating on a scale of 'A' to 'H' where 'A' is the most energy efficient. Nearly all manufacturers quickly, and seemingly painlessly, achieved level 'A' for most of their products so this category was subsequently tightened and labelled as A⁺, A⁺⁺ or A⁺⁺⁺. We all benefit from these regulations.

In 2014, the EU Eco-design regulations restricted the maximum power consumption of household vacuum cleaners to 1600 watts, lowered further to 900 watts in 2017. Newspaper articles misled many people, who rushed to buy older models thinking that more powerful models are more effective. But new well-designed models are quieter and as effective at cleaning. Those who bought old models continue to waste energy and pay more to this day. As a further complication, the manufacturer Dyson successfully appealed against the EU Energy Labelling rules for vacuum cleaners, arguing that the test regime to assess the appropriate 'A' to 'H' label did not take account of real-life usage where the vacuum cleaner gradually fills up with dust. This demonstrates the importance of getting regulations right and kept under review to ensure the optimal outcome.

Other regulations are designed to change behaviour. The Scottish Government is introducing a deposit return scheme to collect and recycle more single use drinks containers and to decrease the amount of litter. Some in the drinks industry have not welcomed this, viewing it solely as a tax on their products. Although it will increase the price of buying drinks, consumers will get their money back assuming they return the container. If, as seems likely, the public return containers using reverse vending - machines that take-in rather than sell products - then this will increase the quality, and value, of materials collected for recycling. Similarly, the UK Government levies an Aggregates Tax, designed to reduce the extraction of rock, sand and gravel and to encourage reuse of waste material from construction and road building. Extracting aggregates does not release carbon (except from the machinery used) but this is still a sensible tax designed to account for some of the other adverse environmental impacts of quarrying, including noise, dust, water pollution and permanent scarring of the landscape.

An example of well intentioned, but poor policy was the decision by the UK Government in 2001 to cut fuel duty on diesel relative to petrol to encourage people to buy more fuel-efficient diesel cars. Unfortunately, diesel cars emitted far more nitrogen oxide and small particulates which increased air pollution. Although this was well known at the time by environmental professionals, it took years for the government to reverse this fuel price differential. In fact, it took the scandal when Volkswagen admitted to manipulating the results of diesel engine tests, to really raise public awareness of this important public health issue.

Planning controls and building regulations affect all construction and urban land-use. Good regulations guide architects, surveyors, engineers and the construction industry towards 'better' behaviour to benefit society. But bad regulations add cost for no benefit, hinder innovation and protect incumbent companies. Buildings must comply with a host of regulations covering health and safety, fire risk, noise abatement, and insulation standards to reduce energy consumption. The Scottish Government has tightened the insulation standards for new buildings in a series of steps. New buildings are far more energy efficient than those built only 20 years ago. Unfortunately, it would seem that both the UK and Scottish Governments were pressurised by industry and house builders to delay regulations for 'zero carbon'– new buildings that would produce as much energy as is required to heat them. The construction sector argued that this would force up costs and make it harder for young people to get on the housing ladder. So, developers still build houses that burn gas for heating and hot water and are not built to the highest insulation standards. This is a decision that we might regret, as it is far more expensive to retrofit houses to the high standard that is needed to meet climate change targets.

Some regulations create a longer-term framework and incentive to change. Several councils in the UK have brought in low emission zones in city centres to combat air pollution from vehicle exhausts. These have an immediate impact on air quality but also, and more importantly, send a long-term signal to car manufacturers and to consumers. Even if drivers do not live in a city affected by such a ban, they start to think that their next car should be electric in case they ever want to drive into a city centre.

Fuel efficiency standards are the single largest action the US Federal Government has taken that has reduced carbon emissions. It was introduced to preserve fuel supplies in the aftermath of the 1973 oil crisis. The standards apply across a manufacturer's fleet. For passenger cars the average miles per gallon (mpg) rose from 15 in 1975 to the new standard of 27 by 1985, later tightened to 35 by 2020. Debate is continuing over how much further to tighten them. In the EU, the average fuel consumption is already over 50 mpg with manufacturers rushing to sell electric vehicles to lower the average emissions of their car fleets.

Sometimes outright bans are required. There is a global ban on the production of CFC's and a planned phase out of other chemicals which destroy the ozone layer. Chemical companies filled the void created by this ban with new innovative products, although in the rush to change, some of these contain gases, such as hydrofluorocarbons (HFCs), which contribute to global warming. These now need to be banned.

There is recent concern about the impact of plastics on the environment. The EU has announced a ban on certain single use plastics such as cutlery, plates, straws and some food containers. A ban is necessary, because even although there are alternatives, consumers keep buying these items as they are cheap and convenient. The Scottish Government is banning plastic stemmed cotton buds used to clean ears. They cost pennies to buy but their inappropriate disposal can cause problems with consumers flushing them down the toilet. The biodegradable cotton is treated at the sewage works, but the plastic stems (like a small straw) are small enough to go through the filters at the sewage works and are discharged to the river then to the sea. I have picked up hundreds of these colourful plastic stems from the beautiful Gullane beach near Edinburgh. Despite the best efforts of volunteers, most will become embedded in the sand and mud, in effect becoming fossilised. In this instance an outright ban on plastic in this product is the best way forward.

Despite what some trade organisations and sections of the press think, we need more well-designed and flexible regulations that will encourage and reward innovation. As new regulations force change, business and consumers can react in unpredictable ways. If the intended impacts are not being achieved, governments need to be agile and quickly adapt regulations. Well-designed regulations will compel or nudge business to design better products to enable individuals to make better carbon choices.

Chapter 9:
Good Design

Design is a difficult concept to explain. For a product, it covers the raw materials that manufacturers use to make it, its size and shape, its energy efficiency in use and the way the manufacturer intends the product to be used. Similar design issues apply to the services that we buy. Design also applies to the places that we all live in – our infrastructure and the built environment. Good design is fundamental to the enjoyment and effectiveness of using a product or service. It is quickly apparent if a product or service is poorly designed, and we all know places that do not contribute to a good quality of life.

Much of the environmental damage caused by a product, service or place is locked in at the design stage. For example, if a car is not designed to be fuel efficient, then it will be difficult or impossible to improve this later. If a housing estate is designed badly and built to a poor standard it will be expensive to upgrade later. Good design influenced by regulations and led by business is therefore fundamental to be able to offer consumers better carbon choices. One essential aspect of good design of products is to optimise their energy efficiency during use.

Products should also be designed using the 'circular economy' principles; as championed by the Ellen MacArthur Foundation. Ellen MacArthur gained the world record for the fastest solo circumnavigation of the globe in 2005. On that voyage she realised that the world is finite and learnt to minimise her use of resources. The three circular economy principles are:
- Design out waste
- Keep products and services in use
- Regenerate natural systems

This chapter will focus on the first of these. Good design is a prerequisite to achieve the second principle. To reduce our consumption of virgin raw materials and our need to continually buy new products the four 'R's are also relevant - Reuse, Repair, Remanufacture and lastly Recycle. Many commentators would insert an extra 'R' before these - to Reduce our consumption.

Products should be designed to be resource effective; that is to minimise the overall use of resources over their lifetime. Minimising the quantity of raw materials required and minimising any waste during manufacture, use or end of life will help. Some designers already do this to a certain extent. But the price of many materials and components does not reflect their environmental damage and are so cheap that the price incentive does not work effectively to reduce our consumption of raw materials. It is possible to reduce the need for raw materials by making things smaller, with less waste in the manufacturing process or by using strong or light materials. For example, the National Grid has built traditional 'A' shaped high-voltage electricity pylons since 1928 with little change in design over this period. Recently they ran a competition to design a more efficient and aesthetically pleasing design. The winning design was a 'T' shape that uses far less steel and is visually less intrusive as its maximum height is 38m versus 50m for traditional pylons.

Sometimes new design can eliminate the need for a product. Yellow Pages, fax machines and public phone boxes were all superseded relatively quickly as new technology replaced and improved on their services. In the past tractors replaced horses to plough fields and automation replaced the need for permanent lighthouse keepers.

If products are designed to last a long time, then this will reduce our need or desire to buy new. The cheapest price often directs or attracts consumers to buying a product, but often this does not benefit us in the long-term. For example, you can buy poor quality magnetic learner driver plates for one pound. They are cheap and convenient, but as soon as the learner driver starts driving faster, or drives on a windy day, the plates fall off and litter our roadsides with plastic and steel. Magnetic plates can work but they need to be designed with stronger, better quality, magnetic material.

Offering a long guarantee when selling a product would incentivise manufacturers to build quality long-lasting products and offers comfort to the consumer. Some new cars come with a seven-year guarantee and a few products, such as Tilley hats and Parker pens are sold with a lifetime guarantee. Clearly it is initially more expensive to pay for this level of quality, but you can guarantee that products will be designed and manufactured to a high standard, reducing the risk to the retailer of returns under the guarantee.

An exception to keeping a product for a long time is if the latest version offers significantly better technical performance or if it is significantly more energy efficient than what it replaces. For example, it is sensible to replace old energy inefficient fridges. This assumes that you do not move the old fridge to your garage just to keep drinks cool!

Products should be designed to be easy to repair or upgrade. Spare parts should be available, and components should be accessible to a person trying to repair the product. Dyson has designed its vacuum cleaners so that it is easy to order spare parts. At the end of a product's useful life it should be possible to refurbish it for resale and reuse. Many large household electrical appliances are difficult to repair because parts have been welded on, making it impossible for a consumer and difficult even for a professional to access any broken parts. Retailers constantly sell new models meaning that the current stock of spare parts will not fit old products. Our whole consumerist society has, over a few short decades, moved from a culture that repaired old clothes and products to a throwaway society where it is more convenient to buy new. However, this is not the case in many low income countries. In the markets of Marrakesh in Morocco for example, tailors repair clothes whilst workshops will repair any product.

Products should also be designed to be easily recyclable, as everything will wear out eventually. If a product is made of composite materials, glued or welded together then it will be difficult to recycle. Similarly, a lot of food packaging, such as disposable coffee cups and some tea bags, look like they are compostable or recyclable, but contain a thin plastic film. Oranges and bananas, in their self-contained compostable 'packaging', have plastic labels stuck on them which are not compostable. To sell to increasingly educated consumers in the future, designers and manufacturers will need to consider how it will be possible to disassemble and recycle the products they sell.

Of course, some items are inherently hard to recycle. Sometimes innovative companies can produce niche solutions. Adidas takes waste plastic collected from oceans and converts it into new trainers. Elvis and Kresse use discarded fire hoses to make wallets. In both cases these products sell at a premium because of the story and provenance of the source materials.

A Scotsman, Graham Tuley, designed plastic tree shelters, or tree guards, in 1979. Tree planters use them to help saplings become established. They act as a barrier against herbivores such as rabbits, voles and deer and protect against strong winds. However, if left too long they can strangle the growing tree, then they disintegrate into thousands of pieces of micro-plastic littering the countryside. Landowners could remove, clean and re-use or recycle them, but as they are often splattered with moss and mud, they are mostly dumped in landfill. Biodegradable alternatives exist, and have been trialled, but none seem to work as effectively as the transparent plastic ones in common use. This problem is not insurmountable. It needs the forestry sector to set up an innovation design competition followed by trials to develop an effective and biodegradable alternative. Tree guards are a visible but minor issue but illustrate a problem that good design and innovation could overcome.

But how many products that we buy fully adhere to these circular economy principles? Have a think about the products that you have bought recently and look at what you place in your waste bin.

****

Over the years working in an office I have lived through several organisational changes. Staff are moved and the office space and meeting rooms are reorganised. Office managers throw out good furniture or donate it to charity as it is no longer needed or is not colour coordinated. But it is possible to design office furniture to be long lasting and to be easy to repair or to refurbish to look like new. This is sometimes easier if the company selling the furniture retains an interest, for example by leasing it to the customer. Buying a service like this, rather than buying goods, can help to support a circular economy. For example, offices usually lease, rather than buy photocopiers. This provides an incentive for the service provider to update and refurbish the photocopier rather than the owner sending it to landfill after a couple of years. New technology is enabling such a service economy. Car sharing websites and apps allow access to a car when you need it rather than having to worry about buying a car, parking it and paying for repairs. In theory this could dramatically reduce the number of cars in the UK with young people in particular never needing to own a car. The RAC estimate that the public only use their cars 4% of the time.

Good design extends beyond the design of products and services. The design of our infrastructure, buildings and townscapes is important to our health, safety and well-being. Cities can be unsightly, congested, polluted and designed to make it difficult to live without access to a car. But we can design cities to be good places to live. Being more compact helps as it avoids the need for a car; alongside attractive buildings, parks, water features, tree lined streets and clean air. Well-designed, clean and accessible public transport with access to live digital timetables encourages people to use public transport.

Green spaces in cities provide multiple benefits. They absorb water, reduce flooding, connect people to nature and provide a habitat for wildlife. They also reduce overheating caused by dark surfaces such as tarmac absorbing the sun's heat. Businesses benefit too. An attractive green city or business park improves health and wellbeing, attracts investment, jobs, people and increases house prices.

Most high-rise offices have their lifts in the lobby near the main entrance. Often these are a central feature of the building and are designed to look attractive. Meanwhile the stairwell is bare-walled concrete consigned to a hidden corner of the building, to function as a fire escape. An alternative approach is to design an attractive staircase in the main lobby that naturally attracts people, with the lift placed out of sight designed primarily to move heavy items. This approach nudges people towards using the stairs which will benefit their health and reduce energy emissions from operating the lift.

Design is fundamental to everything that we do and consume. We should spend more time and effort on design to make the locations where we live more attractive and to ensure the products that we consume are as resource effective as possible. Sometimes good design is immediately cheaper and more efficient than poor design, but often the benefits are long-term or aesthetic rather than price led. Governments should set sensible regulations that business can respond to, whilst sometimes we as consumers should be willing to pay more to buy better designed, higher quality goods, services and infrastructure.

Chapter 10:
Targeted Innovation

Innovation will be essential to reduce emissions and to make us more resilient to the inevitable consequences of climate change. It can make existing technology more efficient or effective, or it can help to reduce prices to enable the mass roll out of a product or service. Innovation can also be a completely new product. For example, DNA analysis of water samples can now quickly identify what species live in a river. This sounds like something from Star Trek, where the starship USS Enterprise could remotely scan a planet for signs of life. But now DNA analysis can be used to monitor fish stocks or to identify new non-native species. Innovation is not always as transformational; it can be incremental, a small but significant improvement such as developing a more streamlined hull on a ship to reduce fuel consumption. And, innovation is not just about new products and services. It can also support new business models or changes in the system. A good example is the circular economy concept of renting products as a service which encourages manufacturers to produce quality and long-lasting products.

Some argue that we can solve climate change with current technology. Certainly, adopting the best technology would cut our emissions significantly, and rolling it out to low income countries is one of the quickest ways to reduce global emissions. However, I doubt that society will tolerate the cost and trauma to achieve zero carbon by just using current technology. Global progress will be far easier and faster with the cost of low carbon options such as renewable energy falling below the price of fossil fuels and if new solutions avoid the public from having to make too many changes to their lifestyles. Whilst we do have effective solutions to most sources of emissions, we do not for aviation and shipping, meat-production or to manufacture concrete and steel. We need to reduce demand and invest in innovation in these areas.

The price of a battery for an electric vehicle fell tenfold between 2010 and 2020 and continues to fall. Samsung has developed a solid-state lithium battery which is safer and twice as energy dense as those made from traditional lithium-ion. New batteries also last longer. The life-time cost of running an electric car will soon be less than a car powered by fossil fuel. And as electric cars can accelerate faster, have more reliable engines and lower maintenance costs, they will become a popular choice even without government intervention. This has only arisen through regulations and consequent investment in innovation by car-manufacturers. Fast chargers and improved energy from batteries are making electric cars a realistic choice for consumers.

Government intervention can steer and accelerate innovation. The EU has long championed innovation directed towards 'public goods' such as environmental improvements. In Scotland, Scottish Enterprise supports innovation. It helps companies to develop a culture of innovation, gives advice on patents, supports collaboration and offers research and development grants.

Big businesses and wealthy philanthropists can also support innovation. There are many innovation prizes available and these can help to focus minds. The Bill and Melinda Gates Foundation is one of the largest. It focuses on innovative solutions for health, agriculture, family planning and sanitation. Such innovation can arise from established companies; but more often originates from new start-ups or from companies working in what seem like unrelated fields. An example is the adoption of LEDs in lighting. The electronics industry initially developed LEDs for mobile phone companies to improve the visibility of screen displays before the technology transferred to providing light to buildings.

Innovation can help to reduce our carbon emissions and support a low carbon economy. Examples abound; including improved design of wind turbine blades, indoor farming to grow local produce all year-round, improved car batteries, greater output from solar panels, cheaper energy storage, and new materials which are lighter, stronger or compostable. The energy sector can also apply digital innovation, for example to integrate renewable energy into the electricity grid or to use weather forecasts to optimise the output from wind farms.

Industry is changing and innovating, faster than most of us can imagine. Industrial biotechnology can be used to make new products, tougher materials are being developed and industry is adopting 'Industry 4.0'. This is the digitisation and automation of manufacturing where sensors embedded in products are connected by the internet and artificial intelligence optimises efficiency. Factories of the future will be automated. Such factories will react to changes in demand, reducing stock and waste. Already it is possible to use augmented reality headsets to train people, or to fix production problems remotely, potentially on the other side of the world. Artificial intelligence can scan and identify objects, for example, to sort mixed plastics on a conveyor belt in a recycling plant, or to identify the correct spare part needed before chartering a helicopter to take it to a remote oil rig.

3D printing is the use of digital instructions to build a three-dimensional solid object layer by layer. It has tremendous potential, some good, some potentially bad. No waste, new efficient designs, no stock required. We could print products locally, in the High Street or even at home eliminating the need to package and transport goods. You will be able to download software instructions to print spare parts making it easier to repair products. But there is a downside. Imagine if you could print as many plastic Lego bricks as you wanted at home. How many would you choose? How many would your children demand? The scope for increased consumption is enormous.

Most chemicals are energy intensive to manufacture but can help to enable a lower carbon society. Examples include insulation, biofuels, lightweight components for vehicles and anti-fouling paint to streamline a ship's hull. Advanced polymers, resins and composite materials have enabled the manufacture of huge blades for offshore wind turbines. When I visited the remote Fair Isle in 1990, a wind turbine produced electricity. It was installed in 1982 with a capacity of 60kW. As recently as 2009 the average offshore wind turbine had a capacity of around 3MW. Now developers are building offshore wind farms in the North Sea, with each turbine having a capacity of 10MW. Siemens Gamesa has announced plans for 14MW wind turbines for the North Sea, with a blade diameter of 222 metres. Each one can power 18,000 homes.

Biotechnology is the use of biological resources to process and produce goods. Bio-based materials are often less toxic and hazardous than synthetic chemicals. Scientists have developed biocatalysts to speed up chemical reactions, at lower or ambient temperatures. For example, new enzymes in detergents enable us to wash our clothes at 30^0C, saving electricity. Bacteria have been found that can clean wastewater, digest waste plastic, remove sulphur from shredded tyres and remove pollutants from soils. However, the carbon and environmental lifecycle of bio-based products is not straightforward and will vary on a case by case basis. Some Coca-Cola bottles now contain 30% ethanol from sugarcane or corn starch but of course this requires land that could otherwise be used to grow food crops or be set-aside for nature. If the sugarcane requires a lot of fertiliser then this could result in high carbon emissions.

****

Innovation can lead to new and difficult choices for society. Humans have selectively bred plants and animals for centuries, but the newer science of genetic engineering is controversial. At best it can make crops more resistant to pests and disease, require less water or nitrogen fertiliser, produce food with greater nutritional value, increase yields and enhance resilience to drought. At worst, it can tie farmers into dependence on a single multi-national company, with the remote but potentially significant risk of genetically modified organisms doing some unexpected harm to the environment.

The UK's universities are world-leading in pure or blue-sky research. Prestige and promotion stem from the number of research papers written and presentations made. However, to tackle the immediate threat of climate change we need to move resources to near to market practical applications that business can then produce or deliver. Some areas of our society are crying out for new innovative ways to be more sustainable. A fully functional waterless toilet would make a significant difference, reducing the volume of treated drinking water used to flush toilets and pollution from wastewater. We could slash carbon emissions if the construction sector developed new ways to substitute our main building materials with wood or hemp, or if we could manufacture bricks, cement and concrete more efficiently. Aviation requires radical solutions as there is no technology on the immediate horizon to decarbonise flights. Unless we are all to stop eating beef and dairy products, we also need to develop new ways to reduce the methane emissions from cattle. Scientists are already exploring selective genetic breeding and trialling, usually at extra cost, new feed additives – bizarrely they discovered that cattle grazing on a diet of seaweed on a remote Orkney island emit less methane.

Meanwhile, think tank RethinkX believes that we are on the cusp of a major disruption to agriculture and food.

Manufacturers can produce proteins in tanks in a factory using precision fermentation. The cost of this new process will fall until it become cheaper than protein from animal-based products. Mixing this protein with other chemicals can produce nutritious foods that may quickly replace protein from industrialised animal farming. This can be produced anywhere in the world, potentially near to market. It uses less land and water and produces less waste and is not prone to disruption from severe weather. The resultant collapse in animal and dairy husbandry could cause the value of farmland to fall and free up pasture and land currently used to grow animal feed. This could lead to an opportunity for massive environmental restoration.

Innovation can enable rapid change, but we need to guide it, and use it carefully if we want to live in a better world. The remarkable development and roll-out of offshore wind technology in the seas around the UK demonstrates that if the right policy, regulatory framework and incentives are in place, then creative and innovative capitalism can quickly deliver transformational change. To cut carbon emissions in the next decade we need to channel our innovation resources into near to market technology, roll-out existing low carbon technology across the world and encourage and enable business to offer us all better Carbon Choices.

Chapter 11:
Targeted Investment

Capital investment is expenditure to replace, upgrade or extend the capacity to produce goods or services, or to build new infrastructure such as roads or a water treatment plant. Such investment needs money to be readily available to pay for the initial cost. Project sponsors and investors need market stability to reduce the risk and have confidence to invest in projects that often have long payback periods.

We need massive and targeted investment to build an economy and society that can support the public to be able to make low carbon choices. We need electric buses, electrified train lines and electric charging points for cars to reduce the impact of our travel. Ultra-insulated homes, district heat networks and heat pumps to heat our homes. Offshore wind turbines, electricity grid upgrades and battery storage to reduce the impact of our electricity use. We need new recycling facilities and carbon capture plants to store carbon dioxide underground. Whilst not all of these investments are cost effective using current economics, most would be if we accounted for the full cost of carbon and environmental damage. Cavity wall and attic insulation are examples of cost-effective investments. Offshore wind turbines are cost effective if you compare them against the full cost of burning coal or gas, whilst electrifying existing rail tracks is desirable but relatively expensive given the need to raise existing bridges or lower the tracks to provide space for new pylons and cables.

Capital investment to replace existing assets benefits the environment if modern equipment, machinery and transport is significantly more energy efficient than what it replaces. For example, Ryanair planned to invest 20 billion euros to buy 210 new airplanes that use 16% less fuel then their existing fleet. This is an example of incremental improvement, a step in the right direction.

An example of 'invest to save' resulting in significant carbon savings is that of Stirling Council, a Scottish local authority. They invested £6 million to fit 37,000 solar panels on 3,900 council owned houses and are now installing 300 batteries to further increase the energy saving to tenants. The government's feed in tariff subsidy pays a private company to install the panels, the council benefits from improving the lives of its citizens and tenants benefit from lower fuel bills. Stirling Council also borrowed £10 million to replace its 12,000 sodium streetlamps with LEDs to reduce energy consumption by 63% and save the taxpayer £30 million over 30 years. Unfortunately, satellite images still show Europe's towns and cities ablaze with the orange glow of inefficient sodium streetlamps. There is still vast potential from energy efficiency that society has not realised.

****

We also need to invest to adapt to the unavoidable impacts of climate change. This provides business opportunities in construction and innovation. In the UK, the most obvious investment needed is in flood management. The predictions are that our climate will have more droughts and more extreme rainfall events. Water companies will need to invest in larger storm drains and sewers to cope with extreme rainfall and at the same time invest in new storage capacity to cope with more intense droughts. Across the country investment is required to reduce the impact of floods, such as planting trees to reduce or slow down water from entering rivers, strengthen protection against landslips and hard barriers to prevent flood water from damaging property and infrastructure. The Thames Barrier alone, which opened in 1982, cost over £500m

- over one billion pounds in today's money. As sea level continues its inexorable rise it will not be financially or physically possible to protect all vulnerable coasts. We will need to invest in sea defences to protect properties that are vulnerable and valuable. Already there is talk about building an even more expensive barrier to protect London. But the obvious and inevitable implication is that we will have to abandon certain areas to rising sea level. Another impact of climate change will be an increase in heatwaves. We will need to invest to upgrade rail tracks to avoid them from buckling in extreme heat and will require new air conditioning in public spaces such as hospitals which are already prone to overheat.

Ski resorts across the Alps have invested heavily in snow making equipment. Initially deployed to enhance snow cover on vulnerable low runs, now they are required on all runs. Snow making is expensive, requires water and ironically uses a lot of electricity. It is like walking against a moving escalator - a constant struggle just to stand still against the increasing effects of climate change. There have even been reports of ski resorts moving snow by helicopter from higher areas to cover bare patches on runs. Adapting to climate change is going to be difficult and expensive.

****

Investment in new capacity to produce goods or services or in infrastructure requires a long-term approach. Governments can make such decisions, but the private sector needs a stable policy environment and financial situation to make investment decisions that will maximise returns and minimise their risk. Unexpected changes in policy and regulations whether brought about by a change in government, government changing tack, or wider political or financial uncertainty all reduce the appetite of the private sector to invest. Investors can delay, cancel or transfer investment overseas to other countries that have a more stable environment.

One example of where the current regulatory and financial market is not working is the failure to build more pump storage hydro-electric power stations in the UK. Pump storage helps to balance the fluctuations between the variable supply of renewable electricity and the demand for electricity by pumping water to an upper reservoir, for future use, when demand for electricity is low. SSE, one of the privately-owned power companies in Scotland, has planning permission to build a new one in the Highlands. This would double the existing pump-storage capacity in the UK and would help to increase the penetration of renewable electricity. Most experts think we need more electricity storage, but the investment case is considered to be too risky because there is no long-term certainty to government energy policy and subsidies.

Deploying innovative technology, even if superior, is always a risk to the private sector. It is often easier to simply build what has gone before and not experiment with the best low carbon option. Examples include the reluctance of developers to install communal district heat in new developments, whilst the aviation industry is notoriously risk adverse in their investment for obvious health and safety reasons. The public sector has a role to play to reduce the risk in deploying such new technology.

Investors and financial institutions fund most major companies and capital investment projects. A question arises as to whether such investors are responsible for the actions of the companies that they invest in or lend to? There is no easy answer but there is increasing pressure and momentum towards such institutions restricting loans to companies that invest in new coal and oil projects. Pension funds are under pressure not to invest their money in such companies or to withdraw investment, known as divestment. Consumer driven social media campaigns can also influence the behaviour of major brands. However, the reality is murky and not straight forward. Lending institutions and shareholders can actively apply pressure on publicly run company boards to influence them. But divestment is the quickest way to lose any influence you have over a company. If pressure is continued to be applied to publicly owned companies, the likely outcome is that they will sell any subsidiaries that operate in fossil fuels and these companies will be bought by private investors, perhaps overseas, who are not subject to the same level of scrutiny. The situation will vary on a case by case basis, dependent on how sensitive the company is to pressure and how easy it is to find alternative sources of finance, or to relocate overseas.

The other side of the coin is to make positive investments in ethical banks such as Triodos or into ethical funds that do good for society. At a smaller scale there are options that enable individuals to invest direct into socially useful projects, by-passing the usual involvement of major financial institutions. The crowd-funding site, Abundance is one player in the market. Individual investors put money into low carbon projects in return for interest payments. At best such schemes offer a higher rate of interest to savers than banks and they bring individuals and communities to work together on low carbon projects. At worst, such schemes encourage small investors to invest in high-risk projects shunned by the mainstream financial community, putting individual investor's money at risk.

As the pressure on governments to act to reduce carbon emissions increases, there is a risk for companies that own long-term assets connected to fossil fuels. These could become stranded like a whale on a beach as the tide recedes. Assets that nobody wants, no longer welcomed by politicians or by society. Most coal fired power stations in the UK already fall into this category. These have become expensive liabilities that their owners have to decommission at further expense along with abandoned deep and open-cast coal mines. Given that we cannot sensibly burn all the coal, oil and gas that we have already discovered, further expenditure on exploring for new fossil fuels is likely to be futile. Across the world many still advocate investment in natural gas infrastructure and gas power stations. It is true that gas is better than coal, but its carbon emissions are still about half that of coal and this is not good enough to meet the emission reductions that we need. Again, there is a risk for investors that their investment will become worthless. Be cautious about any investment relating to natural gas.

We need far more investment targeted to enable us to live satisfying lives but with reduced carbon emissions. Investment in transport, buildings and our energy infrastructure are most crucial. We also need to increase investment to adapt to the now unavoidable impacts of climate change. And a final thought. If climate change occurs as forecast, then essential investment to try and adapt to the ever-increasing effects of a changing climate will quickly become an immense burden on taxpayers and on the entire economies of all countries. In the longer-term, as sea level continues to rise, we will need to relocate essential infrastructure and parts of coastal cities.

Chapter 12:
Education, Jobs and Training

I have one memory from school in the early 1980s about fossil fuels and climate change. My chemistry teacher warned us that society should not waste oil as a fuel to drive cars as oil was running out - we would need every remaining drop to manufacture plastic. At university studying geography, climate change was only on my curriculum because I chose optional modules on Antarctica and on climate change. Even the climate change module treated human induced global warming as just one of several factors that could impact the climate; alongside volcanoes, changes in the Earth's orbit, and cooling caused by aerosols from air pollution. Although the climate does change in response to natural factors, the impact of humans in recent years is the overwhelming influence on our climate. That is not to say that climate scientists were not aware of the effect of carbon dioxide 35 years ago when I was at university, just that it was not the top concern at that time.

More recently, my children were taught about climate change at school, but it was certainly not at the centre of the curriculum. It was mainly taught in geography classes. Geography is the study of the physical earth and human activity connected to it, so it is not surprising that its curriculum includes climate change. However, many children do not choose to study geography after early secondary school so they will have learnt little about climate change at school.

Things are beginning to change for the better. British schools now teach climate change but less on the other critical issue of wildlife, biodiversity loss and ecology. Even today, there are no formal examinations at school on climate change or ecology. Ideally climate change would be embedded within many subjects; as to understand it and its effects requires scientific knowledge of physics, chemistry, biology and engineering; and the social sciences and humanities of geography, philosophy, psychology, politics and history. The starting point is to upgrade the skills of teachers to enable them to impart their knowledge to our children.

Many young adults, keen to join the working world, still graduate from university and college without a strong grounding in climate change and how it will affect the planet, their lives and future career prospects. Good engineering is at the heart of the required transformation to a low carbon economy, but universities have not embedded climate change throughout their curriculum. Rather the student has to opt into climate learning, perhaps in an optional module on renewable energy within a four-year engineering qualification. If nothing else, most things engineers ever design will have to take account of warmer temperatures and more extreme rainfall.

All generations; school leavers, parents, those in work and the retired; simply do not know basic facts about climate change – its causes and its impacts. They hear snippets, often contradictory from television, newspapers, radio, books, social media and from their friends. Then they hear from climate science deniers and others who recklessly or deliberately sow seeds of doubt. There is some uncertainty about the detail of climate change, but the basic facts are well established and have been for some time. How can we expect our society to concur with the changes required if most people do not understand why such changes are necessary?

Workplace training is essential to make progress on climate action. The Royal Scottish Geographical Society has produced a Climate Solutions programme for business and the public sector. Another good example is Yorkshire Water, who run a carbon literacy course for its employees to understand climate impacts on its business, but also to encourage positive action at home. More widely, and using modern technology, it may be easier to train people in carbon literacy than you would imagine. It is straightforward to set up on-line training to reach a massive audience, either to achieve a formal qualification or simply to increase knowledge. Meanwhile social media influencers can reach out to a younger audience.

****

There has been a global shift of population from rural areas to cities and fewer people directly earn their livelihoods from subsistence or commercial farming. Consequently, most adults and children are divorced from farming and nature.

Most children, brought up in cities and spending time at home on electronic devices and computer games, have little in-depth knowledge of the plants, birds and animals that they share their world with. Many children have never visited a farm to see how farmers grow food and keep animals. Traditionally children would have been aware of the life and death of farm animals, but now in high income countries at least, abattoirs are hidden. An occasional school trip or field study is not enough to learn about the incredible role of nature – its wonders, the constant battle of evolution, its importance to humans and its fragility. In fact, many children will learn more about nature from David Attenborough wildlife documentaries than from school.

Being able to name a species is a crucial first step to become acquainted with wildlife but most adults cannot even name all the bird species that share their garden. And, we collectively and quickly forget what is natural. People enjoy the open moorlands, blissfully unaware that trees once covered the Scottish Highland glens. Now Rhododendron ponticum is spreading across these bare hillsides. It is the most invasive non-native plant in the UK. Tourists and many locals welcome the sight of these shrubs with their pretty pink flowers in spring, unaware that they outcompete native wildlife.

If you observe and identify the amazing variety of species that live around you and are aware of the creeping spread of certain invasive species, then you are more likely to campaign to protect nature. But even people who do care about wildlife are often unaware of their impact. Most walkers use a lead as they walk across moors to avoid their dog from harassing sheep but may be unaware that dogs can also harass ground nesting birds.

Meanwhile, the lack of predators in our lives and any danger from wildlife makes us less aware of nature. Contrast this to a bush walk I did in the Okavango Delta in Botswana. There, the guides carry no guns or protection from predators and other potentially dangerous animals including lions, hippopotamuses, snakes and elephants with calves. In the Okavango I felt truly alive, alert and aware of nature in a way that I can never feel in Europe where most of its large predators have been made extinct by humans.

In forest schools, children learn in natural and relaxed settings often, but not exclusively outdoors. The concept originated in Denmark and Sweden. The idea is to enable children to explore and discover by themselves which will enhance their social, emotional and physical development. It engages children in their instinct to be close to nature, to use their senses, observe and explore seasonal changes and to develop curiosity. Fundamentally it aims to build the child's self-esteem.

Although they might sound similar, forest schools should not be confused with outdoor education which is currently popular in high income countries. All outdoor education has a value, but adults usually ask children to do a task like build a den in the woods to encourage team building and achieve an outcome. A risk assessment precedes every activity. In contrast, forest schools encourage self-learning and exploration. In my view, we should encourage people, particularly young children, to be outside more as one small step to tackle the disconnect between people and nature.

Not all education needs to be formal. If we encourage nature within our towns; through ponds, trees, water courses and school gardens; then children will play outside in a rich environment, benefit from these experiences and connect with nature. The problem is how rare this has become in high income countries. Children play indoors or outdoors confined within a fenced off garden, cut-off from real adventure in nature - including playing in woods, ponds, mud and rain.

The UK Department of Transport found that 88% of children aged between seven and ten were accompanied by an adult to school, with nearly half ferried in private cars. The news and social media can exaggerate the risks to children; meanwhile officials feel the need to tighten health and safety rules in response to every incident either due to parental pressure or from the fear of litigation. For example, many schools ban children from throwing snowballs and keep children indoors when it rains or is icy. In the playground children are under constant supervision from adults. Every situation is different, but this approach may backfire with children not learning to manage risk. Being closer to nature may be one solution to help our children develop self-esteem and to reduce the risk of mental health illness. We may try to protect our children from individual harm yet allow climate change to occur with devastating societal wide impacts.

****

The move to a low carbon society has already affected thousands of jobs and many more will be in the future. Colleges have a role to teach the new skills needed as carbon intensive industries decline. In the UK, the number employed in coal mining fell from 1,200,000 in 1920 to around 2,000 in 2015. Initially this was due to new machinery, then imported coal and anti-union policy, and then the industry was finished off by policies to tackle climate change. A similar decline is in prospect for the oil industry although society may need some gas and oil as raw material to make chemicals. For many years, the oil industry has undermined international efforts to tackle climate change. Only in recent years has it begun to come to terms with climate change, initially focusing its efforts to reduce wasteful flaring of methane. Now parts of the industry are calling for carbon capture and storage and for natural gas to be converted into hydrogen - designed to use the expertise built up by the oil and gas industry but also to prolong our use of fossil fuels.

Other jobs may not disappear but will radically change. To tackle and adapt to climate change, farmers and land managers will have to grow different crops and change their farming techniques and land-use practices. Gas boiler engineers can retrain to fit low carbon heat solutions and mechanics will retrain to repair electric cars. Heat pumps to heat our buildings require consultants, designers, heat modellers, equipment manufacturers, installers, and maintenance and control engineers. The electricity grid is becoming ever more complex with numerous sources of small-scale locally generated renewable energy and new demands from electric cars and heat pumps. New tariffs and methods to influence demand will be enabled by smart meters. This brings opportunity to traditional engineers and to electronic engineers, control specialists, data analysts and people with marketing skills.

Opportunities will also grow for foresters, land managers, designers and those in the construction sector building new infrastructure - such as district heat networks, wind turbines, electric rail lines and installing building insulation.

Even those working in marketing, finance, tourism and computing will have opportunities to focus on new growth areas. For example, we need accountants to help companies to make long-term investment decisions to support a low carbon economy. Tourism, and its marketing, may change to focus on domestic tourists and less on the long-distance international market.

****

If the government is to keep the public on board with the scale of the required change, then it should offer support to farmers and businesses that will be disadvantaged. Similarly, people working in carbon intensive sectors will benefit from retraining and learning new skills to help them to find new employment.

Many people do not like change, particularly if forced upon them, but change is necessary and is coming. There may be a gradual shift, or a sudden tipping point when the global public recognises that we must change. To achieve the transition to a low carbon economy we need to encourage the activists, support the silent majority and ensure there are as few people as possible who actively try to block change. Progress will require governments to actively advocate why changes are necessary and to provide financial support to those businesses and people most directly affected. Education and retraining are absolutely essential to ease this process but may not be enough. After all, people continued to smoke regardless of its known dangers. The next chapter explores the emotional and cultural values that can be more important in our decision making than hard facts.

Chapter 13:
Behaviour Change - Consumption

Our carbon intensive lifestyles can be divided into three categories. There are emissions that new technology can tackle with little impact on our behaviour. In the UK, consumers barely noticed the change from burning coal to renewable sources to generate our electricity. And, we will soon all drive electric cars. We will need to change our refuelling habits and there are no gears in an electric car, but fundamentally it does not threaten a major change to our lifestyle. These are the easy emissions to tackle.

The second category is the emissions that inevitably result from the physical infrastructure which society has built and from the financial structure of subsidies and taxes which government places on goods and services. If planners allow out of town shopping centres and industrial estates, then this will inevitably lead to more travel. If the government chooses to build more roads instead of cycle lanes, this will further encourage growth in car traffic. If building efficiency regulations are poor and there is no tax on gas, then this will encourage more use of gas. It is difficult for individual consumers to fight against this without a strong personal sacrifice. In effect the system and the available infrastructure constrain our choices. It takes time, normally decades to change this infrastructure, subsidy and taxation system that society has built up over decades.

The third category is the emissions where our individual choices are important. For example, we can choose what goods we buy, what diet we eat and whether to fly long distances. To some extent we can choose where to live. In these areas, behaviour change is important, particularly that of affluent people worldwide. This chapter focuses on this third category, specifically consumerism. It considers the relationship between economic growth and happiness, explores theories to change our behaviour and gives examples of excess consumption.

Since the industrial revolution, economic growth has been closely correlated with adverse environmental impact on the planet. Access to cheap coal, oil and gas fuelled growth. In high income countries economic growth has meant buying and selling more 'stuff', requiring ever more inputs of raw materials, energy and water. The growth in consumerism has been phenomenal and is remarkably recent. And, consumers easily forget, or are unaware, that their carbon intensive lifestyle destroys habitat for wildlife and causes climate change.

Economic growth has lifted millions out of poverty releasing them from the risk of starvation and disease. It is no wonder that some have a deep-seated concern about giving up fossil fuels. Education will help, but people knew for a long time that smoking was bad for them and for their families, but even this knowledge was not enough to make them change. Our society is addicted to fossil fuel, just as smokers are addicted to nicotine. This suggests that changing people's behaviour is difficult.

A key question is whether economic growth has made people happier - on the assumption that human happiness, or life satisfaction, is a reasonable thing to aim for. This is a tricky question. Undoubtedly economic growth has resulted in better health outcomes and an increase in life expectancy. Those who might hark back to when we all lived in self-sufficient rural communities have clearly forgotten the risk of famine, infectious disease and the common place deaths of mothers and babies in childbirth. However, in wealthy countries happiness is not increasing and many people are unhappy, evidenced by loneliness, anxiety, drug taking, depression and the mental health epidemic. We have moved to a more individualistic, rather than community lifestyle. Many people, especially the elderly, live as couples or alone, often far from their extended family. This puts pressure on young parents who cannot easily find childcare, and places the cost of social care on the taxpayer and government. And it can lead to terrible loneliness.

In high income countries we are obsessed with economic growth, usually measured as gross domestic product or GDP. This measures the total economic activity within a country but includes unproductive activities such as measures to deter crime or clear up after crime. It also enables the overexploitation of fisheries and forests because taking from nature adds to GDP.

Malawi has one of the lowest GDP in the world and surveys show that their overall level of happiness is not high. But from my experience as a director of a small charity that supports projects in Malawi, children there are welcoming and smiling. They are enthusiastic, unlike many British schoolchildren. They are keen to learn as they see it as their passport out of poverty and have aspirations to get 'worthwhile' jobs like being a doctor. They live in simple houses with limited access to consumer goods, yet they play together, work at home alongside their extended families and make toys out of locally available resources. Happiness is not all about buying more stuff. Of course, the adults may not be as happy as those in wealthier countries given their fear of famine and access to affordable healthcare. Is it possible for low income countries to provide their citizens with food, shelter and healthcare without copying our individualistic and consumerist traits? Can high income countries 'export' their strengths without their weaknesses – single-use plastics, fossil fuels, dependence on private cars, individualism and an aspiration for excessive consumerism?

It may be more difficult, but it would be better if societies were driven by measures to increase the wellbeing or happiness of their citizens. In international surveys, Bhutan and Costa Rica score highest in the so called 'happiness index' that measures health, education, life satisfaction and living standards. These countries retain a community spirit and have preserved their forests. Costa Rica is one of the 'greenest' countries in the world and now benefits from eco-tourism. Both countries have stable government and little corruption. Costa Rica has no permanent army leaving more money to spend on the needs of society. All these factors, not driven by wealth and economic growth, result in their high scores.

Happiness is about basic needs; food, water, shelter and health all of which economic growth can help with. After these basic needs have been satisfied, it is less clear whether being rich increases your life satisfaction. What is important is having good friends, being an active part of a community and feeling that there is equity, fairness and opportunity within your society.

****

Change can happen when a small group of activists determine to make a change. This may stall, but sometimes, and often unpredictably, the majority adopt the change and then it quickly becomes anti-social not to. Government can also initiate change. Due to force of habit, campaigns to encourage us stop smoking or to wear seat belts were not successful. It required new laws to force people to change. How many of us want to return to the situation where people do not wear seat belts?

So, we need carbon to be the new smoking. Businesses and individuals should be embarrassed to waste energy, to travel excessively, to eat industrially grown beef and to burn fossil fuels. We need a change in personal values accompanied by a new attitude of 'carbon shame'.

Significant change usually happens slowly. Over one year little changes, but over ten years a lot can. Consider planting a tree. One year later there is little to see, but after ten years there is a maturing tree that attracts birds. Or take the case of Scotland's supply of renewable electricity which rose from 24% of gross electricity consumption to 90% in the ten years to 2019. Few believed this was possible, yet this boom in electricity created mainly from onshore wind turbines, was instigated by a visionary target set by the Scottish Government; supported by favourable UK subsidies, a supportive planning regime and innovation which increased the amount of power each turbine can produce.

Incremental change can suddenly accelerate when a tipping point is reached. This may arise from cost reduction or the spread of ideas through a population. This often happens a bit later than experts expected, but then happens faster than predicted. Mobile phones spread slowly at first. Initially, bought by enthusiastic tech savvy people, we teased those who carried large clunky mobile phones. Then they were adopted for 'emergency purposes', then by the majority and became an essential item owned by over 90% of people in high income countries and 78% in emerging economies. Few could imagine how the mobile phone would change our society and allow us to do new things such as access all information and order our shopping. For the older reader, the plot of the TV programme Blake's 7 was based on the leading characters stealing a supercomputer called Orac which had access to all information. Science-fiction in the 1970s - now a reality.

****

Social sciences have developed theories as to how to influence and change individual and group behaviour. 'Nudge' theory suggests that small changes in the physical environment can influence behaviour. The familiar example is the health led campaign to reduce impulse purchases by removing sweets from supermarket check-outs. Providing detailed nutrition information on packaging may seem obvious, but on average we spend about three seconds selecting an item on a supermarket shelf. Clearly, we do not have time to read all the labels. Changing the product description from 'vegetarian' or 'meat free' to positive language that emphasises the taste, provenance and flavour is more successful at influencing our food choices. However, the UK government's long running 'five pieces of fruit a day' campaign has not succeeded to persuade Britons to eat enough fruit and adopt healthier diets. Public Health England estimates that 60% are still overweight. Clearly nudge theory can shift our choices, although it may not be powerful enough alone to persuade us to make substantial changes.

A picture says a thousand words. When agencies provide people with information on cost savings and even offer free home insulation the take-up rate is low. Showing people infra-red photographs of their homes leaking heat has proved to be far more successful. This picture engages people in home energy efficiency - a concept that is normally invisible and a bit boring for most.

Relatively small changes can ripple out and catalyse a significant change in behaviour. In Scotland, most councils offer free caddies to store food waste in the kitchen before householders empty them into an outside bin for collection. This simple change, at no cost to the consumer, led to a 40% increase in composting food waste. The Scottish Government now plans to introduce a deposit return scheme for drinks containers. Consumers will pay this new 'tax' when they buy goods and retailers will reimburse them when they return the empty container to the shop. This is taking us back decades to when milk was sold in glass bottles and consumers returned them for reuse. Then the advent of cheap plastic introduced the idea of single-use drink containers and plastic bags. It is ironic that the Swedish inventor of the plastic bag considered that a strong, waterproof bag that consumers could reuse many times would be better for the environment than single use paper bags.

Social norms drive learning and behaviour. We observe and often mimic negative and positive behaviour of others. People avoid parking on double yellow lines until they see one person doing it, then copy their (bad) example. Conversely, we are likely to learn from famous role models that we admire; so we can change behaviour by building positive educational messages into popular culture such as theatre, cinema or soap operas.

Peer pressure and doing the socially 'normal' thing that your neighbours are doing is a similar and powerful concept. In California, a study of the spread of roof-top solar panels found that the strongest factor was whether someone else in your street had already installed them. It was not important whether you were a Democrat, or a Republican with a higher likelihood of climate change scepticism. Talking to your neighbours was more powerful than any preconceived attitudes.

There has been, and continue to be, dramatic changes in attitudes to a wide spectrum of our lives - women achieving the vote, wearing seat belts, wearing a tie at work, smoking and drink driving. Thinking about nature, our attitudes have changed dramatically. The Victorians hunted tigers to display and show off the skins in their stately homes, commercial whaling continued until 1986, and it is only recently that a campaign is gathering momentum to ban public displays by killer whales in aquariums.

Changes happening today include the increase in vegetarianism and veganism and a reduction in flying in some countries due to 'flight shaming'. In Sweden, the impact of Greta Thunberg reversed the long-term growth in the number of flights taken. Passenger numbers through Swedish airports fell by 4% in 2019, even before the impact of Covid-19.

However, to enable humans to focus our energy on current survival we have inherited a psychological defence mechanism that enables many of us to ignore long-term existential threats. It is understandable why those who work in the fossil fuel industry might not be keen to tackle climate change. Others have more malignant reasons for climate science denial - they may invest in fossil fuels, dislike state interference or support politicians who campaign against action on climate change. These people build a 'wall' around themselves and are immune to rational argument or education. This is aided as the climate message is psychologically difficult. Until recently most had not really felt or seen the direct impacts of climate change. We can easily think that it will not affect us, it may affect people far away, it will impact in the future, or the science is not clear. Any one of these factors can provide an excuse to discount or ignore the climate message, but all four in combination makes it difficult for society to address climate change. A further factor is the sheer, almost unimaginable scale of the problem - planting a few trees will not be enough to offset our emissions.

We need to exploit the human ability to think and plan, and our interest in leaving a better world for our children and grandchildren. Humans are naturally social beings; we can cooperate for mutual benefit. We just need to learn how to do this on a global scale. Capitalism, using the mechanism of money, has enabled international cooperation between diverse cultures and international trade on an unimaginable scale. Trade benefits both parties, and has dramatically increased economic growth, reducing poverty and disease. The problem is that we emit carbon and destroy nature at little or no cost, resulting in massive over-exploitation. If we could reconfigure capitalism to include a proper price on pollution and environmental damage, then we would be taking a giant leap towards solving climate change.

****

If everyone in the world consumed the same as the average person in Britain then we would need three planets to provide enough food, water and materials to maintain our current lifestyles. We need to consume fewer raw materials and resources than we currently do. Resource efficiency and the circular economy can help towards this goal, but we also need to reduce our consumption, particularly of goods, as it is often easier to provide services with less environmental impact.

Our attitudes to consumption have changed beyond recognition, from the rationing, re-use and recycling practices during the Second World War, to today's frivolous and often excessive consumption. The idea of single use disposable coffee cups or disposable tents bought for festivals would be incomprehensible to our grandparents. As a child, when my friends played 'pass the parcel' at a birthday party all the excitement built up to find out who would open the last layer of (recycled) wrapping paper to receive a small present. Nowadays, parents offer a present in every layer to ensure that everyone receives a present. We are more affluent, buy more and give more but is this always better for our children?

Conspicuous consumption is the buying and consumption of luxuries on a lavish or wasteful scale. Usually the intention is to show off wealth or to enhance prestige amongst friends, neighbours and even strangers. This is not a new phenomenon, confined to high income countries or to the wealthy. Tribal leaders from the Iron Age were buried along with precious items such as swords that the rest of the society could have put to better use, and in Papua and New Guinea indigenous people gathered feathers from the birds of paradise. Today people spend thousands to upgrade their kitchens, and teenagers buy expensive trainers that their parents can barely afford. We make these financial decisions to satisfy a complex mix of desire and need, peer pressure, keeping up with the neighbours or simply to show off.

Amongst the super-rich today, examples of conspicuous consumption are gold watches, diamond jewellery, owning a yacht, access to a private jet or hosting large parties. A much more satisfactory offshoot of conspicuous consumption is the 'Giving Pledge'. This is a campaign for very wealthy individuals to promise to give their wealth to philanthropic causes. This campaign follows in the footsteps of the American, Andrew Carnegie. From humble beginnings in Scotland, he created a fortune in the USA to become the richest man in the world in the early 20th century. He then gave away his entire fortune to worthy causes.

In high income countries, even those of us not in the 'very wealthy' category receive a lot of stuff for free; whether it is business hospitality, presents, incentives to buy a product (a plastic toy with every burger), a T-shirt for every running race we complete. Does the recipient really need and use these products? On holiday, do we really feel obliged to buy poor-quality gifts in a tourist shop to give to our friends and relatives on our return? Napoleon was right - marketing teams have turned the UK into a "Nation of Shopkeepers" and shoppers. We have turned Christmas, Easter, Mother's Day, Father's Day, Halloween, weddings and birthday parties into festivals of consumerism. This has not happened by accident; clever marketing experts, psychologists and advertisers have manipulated us to want to buy more stuff. Parents now buy a disposable costume for their children to wear at Halloween. Many borrow money at Christmas, fearing that they would not be 'good' parents if they cannot provide expensive presents. And people release balloons or Chinese lanterns into the sky only for them to land in farmland or woods, at best littering the countryside, at worst choking animals or causing fires.

The antithesis of sustainable consumption is the fast fashion industry. Fast fashion is when people buy and wear clothes a few times and then dispose of them. Over 100 billion garments are manufactured every year, sometimes in sweatshop conditions - an appalling waste of natural resources and contributing to 10% of global carbon emissions. Fashion designers constantly develop new designs; perhaps just a change in colour, rather than any functional improvement in the product. Marketing teams encourage and persuade the consumer that they need to buy the latest fashion.

Meanwhile, friends of mine managed a year without buying anything new other than food, toiletries, shoes and underwear. Of course, this is not so difficult if you already have excess stuff in your house and easy access to on-line platforms such as E-bay to buy second-hand goods. Other friends moved home temporarily and took a car full of stuff to their new home. Most of their belongings remained at their main home and yet they barely missed them. Many of us, particularly as we grow older, own lots of stuff that we do not really need. The production of all this stuff had an impact on the environment.

It is not clear that buying ever more stuff increases our happiness. We should buy less and buy better quality goods that will last longer. But changing the psychology of shopping will be difficult. Many people receive a 'buzz' from buying stuff. We will need clever marketing to shift people's attitudes and encourage us to invest in long lasting quality goods. Do people need make-up or grooming products to look beautiful or handsome? We need some for personal hygiene purposes, but peer pressure drives much of this consumption, meaning that we can reverse these cultural values.

****

There is a range of possible actions to change our behaviour - direct action through legislation and taxes, subtle shifts in marketing and the use of behavioural social-sciences. Whilst behavioural change is usually slow, often over decades, it can be sudden. The outbreak of Covid-19 has shown how governments can make quick decisive action. Many people resorted to working from home for the first time and international tourism came to a standstill. If we truly treat climate change as an 'emergency' then behaviour can change quickly.

Some people are predisposed to dismiss and reject information that challenges their world view and jump upon any evidence that supports their existing view. They are likely to believe their peers rather than rational argument from politicians or facts from scientists. Social media increases this process as people simply communicate with others with similar values and views. It may prove soul destroying and futile to try and persuade an American on the political right with factual evidence to convince them of the need to act on climate change. Yet it may be possible to work with their cultural values. They might welcome placing solar panels on their roof, attracted by the financial benefit and to gain energy security and independence from big business.

And, a final thought on behavioural change. Many minor changes will make a small difference. We need big changes because the scale of the problem is enormous. In fact, we need to make a small number of big changes **and** many small changes to our lifestyles. People will happily cycle to work or recycle their packaging waste - doing this can make them feel less guilty about their 'big' impacts such as flying. Doing several small 'good' things might be better than nothing, but it does not cancel out doing one big 'bad' thing. We need to consider the relative impact of our actions.

Chapter 14:
Local Communities

So far, this book has focused on government, business and individuals. But another important consideration is the role of community, including indigenous communities that still inhabit vast tracts of our planet. Communities can manage and organise some things better and can help to get more people interested and involved in energy, shopping choices, nature and wildlife.

The concept of indigenous ownership of land feels remote and alien to someone from Europe. Except for the Nordic Sami, there are few lands owned by indigenous groups. Yet, indigenous groups inhabit one-quarter of the world. This includes land in tropical forests, deserts and Arctic regions such as Greenland and Siberia. Around two-thirds of this land is still relatively natural, and 40% is protected in one form or another. Consequently, it includes remnants of important biodiversity.

In the 19th century European settlers in North America caused a devastating disruption to indigenous lifestyles and wildlife. 60 million bison roamed the prairies. North American tribes depended on bison for food, clothes, skins to build shelters and as a religious symbol. Then commercial hunters, with horses and guns, hunted them to near extinction for their skins, to clear land for cattle ranches and sometimes as a deliberate policy to disrupt and force the local population onto reservations. By 1900 there were less than 100 wild bison left. Not only did this devastate the livelihoods of the indigenous people, it also impacted the wider environment. Grazing and trampling by bison created diverse habitat for many species of plants. In contrast, farmed cattle did not follow the migration patterns of bison, ate more of the vegetation and were less resilient to snow and drought. The natural prairies shrunk to less than 0.1% of their former range, but the agriculture which replaced it could not cope with the periodic droughts that are endemic. Across a wide area it is estimated that one-third of the topsoil was lost during dust storms in the early 1900s. More recently there is a movement to restore small patches of prairie, and bison numbers have increased to over 30,000 showing what we can do to restore damage caused by humans.

Generally indigenous people have a strong vested interest to protect their environment. The fact that they rely on ecosystems has preserved areas of tropical rainforest and protected its wildlife. Unfortunately, in many places their lives are becoming more difficult. Warmer temperatures are changing animal migratory routes in the Arctic, drought makes subsistence lifestyles in arid regions more marginal, and loss of habitat, combined with illegal hunting and poaching make it difficult for tribes in rainforests to subsist off the land. Moreover, the western culture of 'development' still encroaches upon indigenous lands. New roads may make it easier for doctors, medicine and education to reach remote communities; but they can also bring disease, logging, mining and agricultural settlers.

The way forward is to recognise the value of indigenous lifestyles and to afford legal protection to their land-rights so that they can choose how their society should develop.

****

International investors and multi-national companies own most tourist resorts. They exploit any area that is scenically attractive and politically open to development. Although this brings jobs and some wealth to the community it can have adverse side effects and benefits to the local community are not always maximised. At worst, seasonal overseas workers take most jobs whilst hotels award large contracts for food and services to distant companies. Many of the purpose-built ski resorts in the French Alps are examples of this type of development. In contrast, in Austria, most accommodation is in small hotels owned by local families. Here, everyone has a vested interest to protect the environment on which wildlife and tourism depends.

It may not be widespread, but tourism can be organised to benefit local communities. In Africa, although foreign companies own many lodges for luxury safaris, local people do own or manage some. This ownership model increases the likelihood of hotels buying food from neighbouring farmers and employing local people. Staff can introduce visitors to local customs, and it keeps money in the economy. In Costa Rica, new hotels and their guests were threatening turtles on beaches where they lay their eggs. Local partnerships now protect the turtles' beach habitat, sustainably harvest eggs for food and allow tourists in small regulated groups to photograph the turtles.

If communities living next to wildlife rich reserves directly benefit from tourism, then they will champion and protect the land and its wildlife. In remote areas, which are hard to police, local communities may be essential to prevent illegal logging of forests whether this is for local firewood or for commercial sale. Poaching is a widespread problem, ranging from elephant ivory for international markets to local consumption of bush-meat from wild animals. Again, if locals benefit from tourists who come to see wildlife, then they will be far less inclined to become involved in poaching and may take an active role to report on and combat poaching – which would otherwise have to be carried out by rangers at great, and possibly unsustainable expense. Local people can also take decisions such as whether to relocate, or even cull, animals during a severe drought and whether to allow trophy hunting.

Trophy hunting is controversial and opposed by many, particularly animal rights activists in high income countries. But assuming it is well regulated, it can bring in a long-term source of sustainable income to help protect wildlife reserves. Involving the local community strengthens the case for whatever difficult decisions might have to be made.

When Guatemala created the Maya Biosphere reserve in 1990, conservationists were disappointed that local communities were given a substantial proportion of the area to manage. In fact, deforestation by illegal cattle ranchers, often supported by drug cartels, is much higher in the government owned protected areas than in the community managed forest. The communities protect the forest as it sustains their livelihoods. They selectively cut trees, certified by the Forest Stewardship Council. They sell nuts and other forest products to international markets through contacts made by the Rainforest Alliance.

Unfortunately, well intentioned interventions by charities headquartered in high income countries have not always managed to protect nature reserves in low income countries. Locals can resent foreign funded rangers employed to conserve nature reserves if they prevent traditional sustainable management of the land. In addition, criminal gangs can expose guards to the threat or attention of those who conduct illegal activities such as hunting for bush-meat, logging and other resource extraction. Every situation is unique, but where possible, involving local or indigenous groups is often the best way to protect nature.

****

Large multi-national energy companies lead the energy revolution that is sweeping the UK. But there is a small, but locally significant, community energy sector. When the rush to erect wind farms was at its peak, activists in the village of

Fintry were supported by Local Energy Scotland to negotiate with the multi-national developer. The result is that the community owns one of the 15 turbines erected. This far sighted decision brings a steady flow of income into the village. This is used to refurbish community owned buildings, give local energy advice to householders and to install home insulation. The wind farm developer benefitted from engaging with a supportive community rather than one actively campaigning against their plans. But community energy requires time, money, effort and patience, and is not always successful. Its growth has been hampered by constant changes in the available subsidies. Investments also come with a degree of financial risk and this can cause disagreement within communities.

A small charity, Greenspace Scotland, campaigns for councils and the public to value the open spaces scattered across every city and town in Scotland. Many parks and publicly owned open spaces are sadly neglected after years of declining council budgets. Greenspace Scotland argues that these parks are suitable to generate renewable power, particularly ground source heat pumps to heat surrounding properties. An example is Saughton Park in Edinburgh where the council, with the support of 'Friends of Saughton Park' installed a small hydro-electric power station in the river. The electricity powers heat pumps connected to boreholes under a car park and to ground loop heat collectors installed under a football pitch. The electricity is also used by the park café with any surplus exported to the grid providing a sustainable income which is used to maintain the park.

Government and cash strapped councils are never likely to satisfactorily deliver all activities and services. Community litter picks, beach cleans and removing non-native species are examples where local people have a long-term strategic interest in doing the job properly. Everyone should be encouraged to pick up litter or clean beaches. This could transform our attitudes to cigarette butts, flushing inappropriate items down toilets, crisp packets, plastic bags and bottles, polystyrene packaging and marine litter in general.

Communities can be involved in all sorts of local activities, dependent on their interests and ability. There are urban farms, producing food in the heart of cities, and a campaign to grow community orchards, with the fruit available for anyone to pick. There are many community owned woodlands within and surrounding towns, usually for amenity purposes rather than for commercial gain. Young parents have adopted play parks. Clearly parents have an interest to ensure their local play park is safe, clean and free from vandalism, and as a result often do a better, more reactive, job than the local council.

Some communities and charities have successfully set up organisations to repair, reuse and recycle products that would be uneconomic for commercial companies. People can donate old bikes to workshops where volunteers or people on training placements can strip them down, refurbish and sell. Volunteers have established repair workshops where the public can hand in items for repair. The volunteers gain friendship, and satisfaction from using the workshop tools to give a product a new lease of life. Many offer work experience to people with learning difficulties or to those with a criminal record.

A variety of other activities are best organised by community-based organisations. Parents can organise to pass children's toys, clothes and school uniforms to parents of younger children to avoid clothes going to landfill. Organisations can store and share garden equipment, sports equipment and tools. Charity shops have a valuable role to raise funds and to ensure a second life for goods.

A new shop, Weigh Ahead, has recently opened in my local High Street. Its aim is to eliminate the need for packaging, particularly single-use plastics. The shop weighs the customer's own containers, fills them with goods such as rice or pasta, and then re-weighs them to calculate the amount due. A charity runs the shop and local people contributed to its start-up costs through a crowd funding campaign. Again, the aim is to provide a socially beneficial service. Businesses can also organise themselves as a community. Business Improvement Districts are schemes where businesses in a geographic area agree to pay a voluntary contribution that is spent to enhance the business environment and shopper experience.

Unfortunately, in the UK, change, often due to new on-line services is ripping the heart out of many communities, particularly rural ones. Local services are closing including grocers, banks, post offices pubs and schools. On the other hand, technology can enable new service models such as access to remote doctor appointments and aggregating orders for community food deliveries, whilst previously remote public agencies can consult on-line with local communities to develop new policies.

The Leven Partnership consists of public, private and community organisations working together to bring the River Leven back to life and to reconnect people with nature. The Leven is a short river in Fife, Scotland, which flows through an area of industrial decline. There are spoil heaps from mining and derelict paper mills along its banks, part of the river is underground in culverts and there are few footpaths making the riverbank inaccessible. In places, landowners have straightened and canalised the river to improve drainage, whilst its surroundings are prone to litter, vandalism and invasion by non-native plants. In short, the river is not an attractive asset to the community. Its waters are clean, but it is a sadly neglected river. Scottish Water has now installed a new sewer pipe to replace one that was leaking raw sewage into the river. New footbridges and footpaths are being built, volunteers are tackling non-native species and a derelict railway line and station will be reopened. Fife College is teaching its students about the approach to this successful partnership - involving the community to turn an eyesore into an asset to stimulate wider social and economic regeneration.

****

Community involvement can also be harnessed at a national level. The 2050 Climate Group was established to mobilise young people in Scotland to take action on climate change. It is a volunteer led charity that runs a climate leadership development programme and encourages its members to engage with policy developments and promote systemic change.

The UK Government is also trialling an approach to develop policy to meet its net zero target. A random cross-section of the British public is taking part in a Climate Assembly. The group met to listen to expert speakers and discuss the options around transport, buildings, agriculture and behaviour change. This community-led approach has provided a snapshot of informed public opinion on these complex matters and will report this to parliament to influence future policy.

Communities have a role to help cut our carbon emissions, to preserve wildlife rich areas and to bring us all closer to nature. Communities can sometimes collaborate with people and business in a way that central or local government finds difficult. But, to thrive, community groups often need a little help, whether this is to access finance or simply to operate within a supportive environment created by government and their local authority.

Chapter 15:
Resource Efficiency

By investing in resource efficiency, you get more useful product or service, from fewer materials with less waste. For example, an energy-efficient boiler in a factory can produce more products for less energy, or a new strain of wheat will increase yield as inputs such as fertiliser stay the same or decrease. Contrast this with an increase in food production due to an increase in fertiliser use – more output from more inputs – but not resource efficiency. Investing in resource efficiency reduces the energy required to extract and process materials. The focus here is on energy efficiency but investing in resource efficiency can also reduce water and materials use and associated carbon emissions. Other chapters on design and on business cover the resource efficiency benefits of a circular economy in more detail.

Investing in energy efficiency is one of the most cost-effective ways to reduce our carbon emissions. Yet, it is a choice individuals and businesses need to make, and many choose to spend their money on a foreign holiday or to build a bigger factory to sell more product. Unlike some investments that might be necessary to reduce our emissions, most energy efficiency investments have a positive return to the investor or consumer. Usually there is an upfront capital cost, such as buying loft insulation, followed by years of cost savings. The payback period is the number of months or years it takes before the savings exceed the initial investment. Energy efficiency can be marketed as 'invest to save', and such schemes create jobs and are good for the economy. But, interestingly, many companies are happy to invest in projects that aim to increase sales but are less willing to invest in energy and resource efficiency even if these are more profitable. It is often seen as more ambitious and exciting to increase sales rather than increase profits through reducing overheads.

Energy efficient companies will have lower overheads, making them more competitive in international markets, and energy efficient products are beneficial for consumers. So, energy efficiency should be a 'no brainer'. Yet consumers and businesses do not always take it up to its full extent. There are many barriers to invest in energy efficiency, some apply to companies, some to consumers, but most apply in one shape or other to both. Business or consumers might simply be unaware of the better energy efficiency of one product or service over another or they might be confused by lack of information or conflicting claims made by retailers. Consumers might not understand the new technology or not want to take the risk and may not have time to research the market. Energy efficiency may not feature as a priority when choosing a product and the payback period might be too long for the company or consumer. Last, but not least, the extra up-front cost might be a barrier.

As a result of these well-known barriers, government agencies run active programmes to promote energy efficiency to business and consumers. Most provide information to help consumers to make better decisions and offer loans or grants. These programmes always have some positive impact but rarely as much as was hoped for. Despite support from government agencies many companies do not fully invest in cost effective energy efficiency measures and many consumers do not even accept the offer of free home insulation. All the barriers need to be overcome before someone will invest their time, effort and money into energy efficiency. There seems to be a strong behavioural aspect to change, more fully explored in the chapters on behaviour change and business, consequently governments should also introduce regulations to force change to maximise the benefit to society of energy efficiency.

****

The EU identified that household appliances were major users of electricity. Manufacturers were not taking up opportunities to design energy efficient products, retailers were marketing products based on the latest settings and gadgets, and consumers were ignorant of how much these products cost to run. Selecting a new fridge-freezer can be a daunting task as there are so many models and different features. In 1995, the EU introduced a new labelling system, that tests and rates washing machines, fridges and freezers on a scale of 'A' to' F', with 'A' being the most energy efficient. Very quickly most met the 'A' energy rating so the EU introduced tougher categories. But the rating is based on the energy consumption per area of cooling space, so a large 'A' rated fridge-freezer will still consume quite a lot of electricity. Retailers now promote large 'A' rated American style fridge-freezers with double opening doors with permanent iced water on tap – then homeowners may redesign or extend their kitchen to find space for one. This highlights that we need to consider saving energy in its wider context.

LED light bulbs have been another success story, slashing energy consumption at home and in offices. Energy consumption can be cut by a remarkable 90% when combined with smart controls that automatically switch lights off in an unoccupied area of a building. India is rapidly rolling out access to electricity to its rural population. Light bulbs consume one quarter of global electricity, and this encouraged the Indian Government to act to stimulate the take up of energy efficient LED lights. They placed orders with China for large quantities of low-cost light bulbs, reducing the price from 400 rupees to 70. The number of light bulbs sold across India increased from five million in 2014 to 670 million in 2018, and the market share of LEDs increased from 1% to 46% during this period. This reduced the growth in energy consumption by the equivalent power that Denmark uses, albeit many of these bulbs will need to be replaced because of their poor quality.

It is now possible to build offices and homes to such a high standard that they require virtually no direct heating even in a European winter. Our body heat (around 100 watts) and waste heat from computers and electrical equipment is sufficient to keep the temperature comfortable in all but the coldest weather. Good air flow is important to prevent condensation, and this requires a ventilation and heat recovery system. This extracts warm air from the building, 'captures' and returns it. The premises do not require a boiler for space heating, although hot water will still require a heat source.

In industry and commerce there is enormous potential to further improve energy efficiency. This includes electric machines and devices such as motors – they are often run sub-optimally at full power when not all the capacity is needed. Variable speed drives may cost more to install and are more complex to operate but will quickly pay for themselves. Power stations, electrical devices and heating boilers release huge amounts of waste heat direct to the atmosphere or into water bodies.

Fridges and freezers use a lot of electricity to cool products. To cool a space, you extract some of the heat and expel it. Some supermarkets capture this waste heat and use it to heat adjoining office space; but most 'dump' this heat into the supermarket or expel it through an outside vent to the atmosphere. If the waste heat is released into the supermarket in summer, the air conditioning might have to work harder to cool the shop down. A similar effect used to occur in many offices where heat from computers, monitors and printers would overheat offices in summer. This issue has decreased with the advent of energy efficient equipment which releases far less waste heat. Moreover, laptops use less energy than desktop computers and tablets and mobile phones use even less. These portable items are smaller, and are designed to be energy efficient, not primarily to reduce our electricity consumption, but to enable a longer battery life.

Not only does refrigeration consume a lot of power, most use refrigerant chemicals, such as HFCs, that are powerful greenhouse gases if they leak into the atmosphere. Mackie's dairy farm in north-east Scotland invested in new technology to tackle both these issues. A biomass boiler replaced the previous oil boiler to produce steam to pasteurise the milk. Ammonia is heated with some of this heat. Like HFC's, as ammonia changes from liquid to gas it takes heat out and cools the surrounding space. Unlike HFC's, ammonia is not a greenhouse gas.

Lord Kelvin, who studied at Glasgow University, was the first to propose the concept of heat pumps to heat buildings in 1852. Heat pumps use energy to reverse the normal flow of heat from hot to cold and therefore are sometimes described as a fridge in reverse. Electricity is used to efficiently warm up existing sources of heat using compression. For every unit of energy input, a heat pump can produce two or three units of heat since they extract heat from the surrounding air, ground or water. Even more units can be produced if the source of heat is warmer to begin with, so heat pumps are ideal in scenarios with access to warm ambient water or pre-heated water such as the wastewater from industrial processes and even water in sewage pipes.

A pioneering green heat network has been built in Stirling, Scotland. It uses wastewater entering the sewage works integrated with gas from the digestion of sewage sludge. Heat pumps boost the heat from the relatively warm wastewater to a useable temperature then the hot water is distributed through insulated pipes to heat a leisure centre, school, sports stadium and offices. Heat pumps can also use water from rivers or the sea and can extract heat from the air – more about this in the chapter on our buildings.

****

If water utilities could more thoroughly adopt nature-based solutions, then this would be more cost efficient and reduce their environmental impact. Water is a heavy material - it can require a lot of energy to pump it to our taps. A cocktail of single-use chemicals is used to purify drinking water, but it is difficult to remove all pesticides and nitrates. It can be more cost effective for a water utility to collaborate with, or even pay, farmers to reduce their use of fertiliser and pesticide or to apply them at a different time dependent on the weather. Also reducing customer demand, for example, using water efficient appliances and reducing water wasted from leaks in pipes helps to reduce the environmental impact of our water consumption.

In high income countries, water utilities collect wastewater at sewage works, mix and aerate it to enhance bacterial growth and add chemicals to treat it. This is an energy intensive biological process that also emits methane and nitrous oxide. Some utilities recover energy from the sewage sludge using anaerobic digestion to create biogas with any remaining sludge used as a fertiliser. But on average only a fraction of the energy embedded within the wastewater is captured for reuse. Already, it is common for wastewater to be cleaned so it can be used downstream for irrigation purposes, but Singapore and Windhoek in Namibia have proved it is possible to clean it back to drinking water standards using current technology.

In future, using processes more akin to those in nature could further transform water treatment. Instead of relying on vast volumes of bacteria to break down nutrients and organic matter, it will be possible to select specific microbes and new biological enzymes to speed up chemical reactions. These processes will be more effective, efficient and require less space. Alternatively, wastewater could be treated in an environment with no oxygen to avoid the energy intensive aeration process. Effective treatment of wastewater can also produce a fertiliser from nutrients such as phosphorous and nitrogen, recover cellulose from toilet paper and recover heavy metals, such as platinum particles from catalytic converters which end up in road drains and sewers. A further technique is to replace chemicals with biological processes such as worms or flies. As larvae, black soldier flies eat twice their bodyweight every day, sanitising waste in the process and producing a fertiliser, an animal feed and fats that can be used as biofuel. Another by-product is chitin, a chemical which the utility could use as a flocculant in water treatment. So, wastewater treatment will be transformed from an energy intensive process to one that generates more energy and where chemicals and nutrients are recycled.

Most manufacturing processes use water. Levi Strauss calculated that each pair of jeans requires over 2,500 litres of water over its lifetime to grow cotton, manufacture and wash clothes at home. They have introduced several water saving techniques to process cotton fibres, manufacture and dye jeans. Waterless ozone technology has replaced the optional and water intensive process of stonewashing that makes jeans look worn before sale. Reducing water use and recycling water within manufacturing plants also reduces the chemicals used to treat water, energy to pump it to the plant, energy to heat water within the plant and the volume of wastewater to be treated. Despite these dramatic improvements in water productivity in the manufacturing process, the largest potential water reduction technique in the lifecycle of jeans is to use recycled cotton rather than virgin cotton grown on irrigated land.

****

Apart from our buildings where we can virtually eliminate the need for heating, there are physical limits to energy efficiency. Boiling a kettle of water will always require a certain amount of energy, and some energy is required to carry goods and people from one place to another.

There is another important factor that limits energy efficiency. The 'rebound effect' reduces the actual savings from investing in energy efficiency. Like a boomerang that you throw, but it comes back to hit you, the energy savings you think you will make do not always materialise. If you make things more energy efficient, this will have knock-on impacts which may increase energy consumption. Recalling the earlier example that coal consumption increased after the introduction of more efficient steam engines, some argue that the rebound effect does not result in a reduction in total global energy demand. Instead it makes the economy more efficient, leading to more growth and more energy consumption. Any fossil fuel that we 'save', reduces their price, and this may result in increased consumption of fossil fuels overseas. On a smaller scale, people have replaced single light bulbs with multiple energy efficient LEDs, or string hundreds of bulbs from their Christmas trees – indoors and outdoors. Similarly, some who buy a new energy efficient fridge place the old one in their garage to keep drinks cool for the occasional party, whilst in our homes, better insulation could reduce heating costs or people could simply enjoy the comfort of living in a warmer home. And, if you choose to use less energy and save money, you will be likely to spend that money on some other activity which will lead to more economic growth and probably further energy use.

This is a startling claim, and it is certainly true, that with population growth and despite massive resource and energy efficiency improvements, the total global consumption of energy keeps on increasing. This suggests that to reduce global carbon emissions energy efficiency is best delivered alongside a high price or cap on our total energy consumption. Despite these doubts, resource and energy efficiency is still a net benefit to our society. It allows us to do more with less and to enjoy a higher standard of living.

Section Three:
Solutions – Applying the Building Blocks to our Lives

Introducing Five
Common-sense Principles

This section introduces five common-sense principles which government, business and consumers can use as a guide to make better decisions. Each of the following chapters then looks at a different aspect of our day to day lives and links them to one or more of these five principles. Applying these common-sense principles, will simplify what at first seems a daunting task, so that we can all make better decisions for the climate and nature.

Five common sense principles:

1. **Be fair across current and future generations**
2. **Price carbon pollution**
3. **Consume carefully, travel wisely**
4. **Embrace efficiency, avoid waste**
5. **Nurture nature**

There is some overlap, but the first two are primarily relevant to governments and businesses making strategic decisions, whilst the others are more relevant to us all in making our daily choices.

'Be fair across current and future generations' combines the ideas of equity and social justice. Equity applies between people who live in different countries, between rich and poor, those from different ethnic backgrounds and gender within countries, and between the young, old and future generations. Flowing from this are the concepts that we should take care not to penalise the less wealthy with tax rises, wealthier countries that caused climate change have a duty to help other countries, and we should not leave a toxic legacy to future generations.

'Price carbon pollution' is the concept that we should account for the external costs caused by our emissions of carbon dioxide, methane, nitrous oxide and other greenhouse gases. Carbon and other pollutants cause damage that falls on wider society and wildlife, with no economic incentive for an individual or company to reduce their emissions. Governments need to overcome this failure of the free market through taxes, subsidies, carbon trading, carbon offsetting or through bans and strict regulations.

'Consume carefully, travel wisely' is about the choices we make as businesses and individuals. Businesses can adopt circular economy practices. Consumers can buy and consume less and buy second-hand. Pay for quality goods that will last longer. Buying smaller can also make a significant impact, whether it is a smaller car, house or fridge; or using a smart phone instead of a desktop computer. Consider the need for travel, share travel and make low-carbon travel choices.

'Embrace efficiency, avoid waste' is a no brainer for business and consumers as it combines environmental benefits with cost savings. Energy and resource efficiency will reduce our overall requirement to extract resources from the environment, minimise pollution and cause less disruption to nature.

'Nurture nature' means to take care of and protect. We should restore, enhance and rewild nature. Individual consumers can apply pressure on business. Business can influence their supply chains, particularly when buying commodities from overseas. At home, children and adults need to rediscover, learn, and enjoy nature. We should rewild our lives and the countryside as wildlife needs space to thrive.

By thinking about and applying these five principles; governments, business and individuals can start to reduce their impact on climate change and reverse the loss of wildlife. Governments could apply these principles to every new strategy, policy, law and tax; businesses to every investment decision; and individuals to the decisions and purchases we make. Consumers can also apply them to pressure governments and business to change. These five principles provide a framework for us to all make sensible, common-sense decisions in our lives.

Chapter 16:
Generating Electricity

Section three begins with how we generate our electricity as many commentators believe electricity will become even more important to our society than it currently is. Around one quarter of global carbon emissions arise from generating electricity. The global demand for electricity is rising. Low income countries are adopting lifestyles that require more electricity, whilst high income countries are slowly electrifying their economies with electric cars and heat pumps to heat homes. Governments and the energy companies have the main responsibility to implement the recommendations in this chapter; as at present, individuals and businesses have a limited, although growing, choice over our electricity supply and demand. Later chapters will focus on areas of our lives where we do have more control and choice. Of my five common-sense principles, 'price carbon pollution' is the most applicable to this chapter followed by 'nurture nature'.

Energy utilities invested three trillion US dollars in new renewable sources of electricity in the last decade, increasing the share of global electricity generated by renewables to one-third. For new electricity capacity, renewable sources increased from 25% in 2001 to 75% in 2019. Much of this recent investment supplies additional capacity to meet the increase in global demand for electricity rather than to replace fossil fuels. So, despite this progress, there is still a long way to go and investment in renewable energy needs to increase further and faster.

Our electricity is generated from coal, oil, gas, nuclear and renewable sources. Burning coal emits more carbon dioxide per unit of energy produced than all other commonly used fuels. In addition, methane can leak from working and abandoned coal mines, mining coal is disruptive to the landscape and often dangerous for its workers, it creates air pollution, and is inefficient as up to 70% of the energy produced can be lost up the chimney as waste heat - although combined heat and power coal stations can capture some of this for industrial use or to heat a district heat network. Even so, an immediate priority is to stop building new coal power stations, then to phase out existing ones. Leading the way is the UK, the birthplace of the coal-led industrial revolution, which is amid a dramatic and rapid process of closing its coal power stations with all due to close by 2024. In 1990, coal generated 75% of the UK's electricity; in 2019 it was only 2%. This is a remarkable achievement driven by market forces manipulated by the government who introduced a price of around £18 per tonne on the carbon emitted by generating electricity through the Carbon Price Floor scheme. Around the world, coal fired power stations will become increasingly redundant as they lose their social licence to operate and the falling price of renewable alternatives undercut them. Governments should accelerate this by taxing electricity production directly proportional to its carbon emissions. As this phase out gathers momentum, the world is beginning to appreciate the principle of 'unburnable coal', that is known coal deposits that will never be extracted. The implication is that there is little point in prospecting for new coal reserves and investors should be wary of extracting, processing, transport or burning of coal. If possible, those working in the coal industry should consider relocation or retrain as there will not be a long-term future in burning coal.

Some argue that 'clean coal' has a future – presumably, coal with carbon dioxide and the gases that cause acid rain safely and cost effectively removed. A few years ago, there was a proposal to capture the carbon dioxide from Scotland's largest coal power station at Longannet; using chemicals to capture the carbon dioxide, compress it, pump it through pipelines and store it deep within disused oil fields hundreds of kilometres away in the North Sea. The cost of this was huge, with little direct benefit to the UK. The energy required to operate the carbon capture plant would have increased the total fuel consumed by around 20%. Furthermore, at best, it would have only captured 90% of the emissions from both the normal consumption and from the extra 20% required. So, the proposal would have burnt more coal and only partly negate the negative environmental consequences of using coal - more cost for little direct benefit to consumers. Fortunately, the UK Government rejected this proposal and subsequently Longannet power station closed under pressure from carbon taxes.

Nuclear power emits no carbon dioxide when generating electricity. But it cannot claim to be completely carbon free as the latest nuclear power station in the UK involves a massive £20 billion construction process and mining uranium will have an ongoing environmental impact. Thousands of tonnes of rock are extracted to produce one tonne of uranium fuel, often leaving a toxic legacy of contaminated water and low-level radioactive waste. Moreover, nuclear power, with its legacy of radioactive waste, clearly fails the principle of considering the long-term. I suspect that our descendants will not thank us for leaving them our nuclear waste to deal with. Still, dealing with the waste is not insurmountable, and given its low carbon emissions, each country needs to decide whether they want nuclear power depending on public acceptability, cost, their need for a large low carbon base-load and what alternatives they have access to. While an American political advisor once said nuclear power would be "too cheap to meter", it has never achieved its promoters' dream of producing low-cost electricity because of the engineering required to address safety fears, the constant changes in design and the fact that reactors are mainly built on a one-off basis rather than mass produced. In fact, the subsidy given to build the UK's latest nuclear power station is greater than that provided to generate electricity from new offshore wind farms in the North Sea. However, we do want to avoid what happened in Japan, when carbon emissions increased when they temporarily closed all their nuclear power stations after the Fukushima accident.

Electricity generated from natural gas produces around half the emissions from coal, although there is an additional risk of methane leaking into the atmosphere when extracting and transporting gas. These leaks can be difficult to identify and measure, and at worst could make burning gas nearly as bad as burning coal. Even well-managed gas production combined with an efficient combined heat and power plant emits carbon dioxide, and it is of concern that gas is still promoted by many as a low carbon fuel or a 'transition' fuel until renewable power replaces it. Power stations are built for at least a 25-year lifespan, so building a gas power plant locks us into an unsustainable path. Of course, burning natural gas at scale, for example at a power station, makes it possible to capture the carbon dioxide and store it safely underground. Given that gas produces half the emissions of coal, this might be a more viable option than capturing carbon dioxide from a coal power plant, but government regulation and subsidy will be required to persuade energy companies to capture carbon dioxide from their gas fired power stations.

Unless some new form of electricity generation, such as small-scale nuclear fission reactors or fusion - joining two atoms together rather than the more inherently dangerous process of splitting atoms in nuclear fission - is developed, we ultimately need to source most of our electricity from renewable sources. These include onshore wind, offshore wind, solar photovoltaics, concentrated solar furnaces, hydro-electric, burning biomass, geothermal, wave and tidal.

The best solutions will vary depending on the local geography; for example, countries with deserts and those close to the equator will benefit the most from cheap solar energy; coastal countries in northern latitudes from wind; whilst countries like Iceland, with magma relatively near the surface, can cost effectively drill for geothermal power. One scenario is for an integrated pan-European renewable electricity grid combining solar from Mediterranean countries and North Africa, with wind from north-west Europe, pumped storage hydro-electric from Scandinavia and possibly an inter-connector to Iceland for geothermal energy.

Even renewable energy is not a guilt-free passport to consume as much electricity as you want. All sources of electricity have some adverse environmental impact. Hydro-electric power from dams requires concrete, can damage river hydrology, wipe out migratory fish, and rotting flooded vegetation can emit methane. In addition, large dams can be a source of conflict between countries that share a river basin. Wind turbines require steel, their blades can kill birds, they can be noisy, and some people feel they are unsightly. Solar panels require extensive mining to gather rare metals and their mass deployment is likely to disrupt desert environments. Burning biomass is controversial as it creates air pollution and it is likely to be sourced from monoculture forest with limited habitat for wildlife. Moreover, scientists cannot agree on the net carbon benefits of burning biomass which will depend on the type of biomass, where it is sourced, how fast forests might regrow and how far it is transported. And, growing crops or timber for heat, electricity or fuel on land that might otherwise be used for agriculture can push up the cost of food.

In addition, the quantity of electricity produced from renewable sources is variable, highly dependent on factors such as the weather. To cope with this may require more infrastructure including storage. Then there is sulphur hexafluoride (SF6) – possibly the most dangerous gas that you have never heard of. Each molecule of this human made gas has a global warming potential of 24,000 times that of carbon dioxide - the highest of any known gas. It is a highly effective insulating material used by the electricity industry to prevent short circuits and accidents. Nike trainers used to use the gas as air cushioning for running! Whilst such frivolous uses have ceased, its global use is increasing as utilities use it in wind turbines and in switchgear built to manage renewable electricity. There are alternatives for medium voltage applications, and Scottish Power Renewables is using these in individual turbines for the new East Anglia offshore wind farm. Meanwhile Scottish and Southern Electricity Networks has installed innovative circuit breakers at its high voltage substation at Dunbeath in Scotland. But, not surprisingly the industry has to balance the lowest cost, reliable, efficient, risk free methods with the environmentally optimal solution, and will only rarely risk experimenting with new materials unless incentivised to do so. In the meantime, pending the further innovation and restrictions which are needed, the focus must be to prevent leaks and to recycle SF6 effectively when the equipment is replaced.

Despite these concerns, using renewable sources is the best way to a clean and sustainable future. To decarbonise global electricity will need favourable planning rules, further innovation, subsidies and regulations designed to maximise the generation of renewable electricity. Despite frequent policy changes, the UK is a good example of how to transform an electricity grid from dependency on fossil fuels to a low carbon mix of renewables and nuclear – but care is needed through this transformation to ensure grid stability. Given the privatised energy market in the UK, this transformation required government policy and active intervention to steer energy companies towards the desired outcome. Placing a price on carbon emissions to cover its environmental damage led to the demise of coal, whilst government led 'contracts for difference' reduce the cost to the taxpayer - private companies bid into competitive auctions to supply an agreed amount of renewable electricity. However, every country needs to design their own policy and regulations to suit their different starting points, regulatory regimes and ease of access to low carbon sources of power. In the UK, the most obvious expansion is to build new onshore wind turbines and vast new offshore wind farms. With new floating turbine technology, it should be possible to build further offshore into deeper waters often with more reliable wind speeds.

Planning rules by local councils restrict the deployment of renewable generation in certain places. Many will not want to see intrusive solar panels on roofs in a conservation village, but it should be possible for every farm in the country to install solar panels on farm buildings and to erect a medium sized wind turbine, yet many councils ban wind turbines across wide areas of the countryside. When planning consents for existing large-scale wind farms expire, after say 25 years, we need a planning regime that will look favourably on well-designed proposals to install larger, more powerful wind turbines on the same sites. Scottish Power built their first commercial wind farm in Scotland in 1995 at Hagshaw in Lanarkshire. They erected 26 turbines with a total capacity of 16MW. In an illustration of the amazing rate of technological improvement in renewable energy, they are replacing these with 14 larger, more powerful 6MW turbines with a total capacity of 84MW and are installing an on-site 20MW battery storage facility.

Feed in tariffs are a subsidy paid to consumers in the UK to generate their own renewable electricity. Governments use these to kick-start new industries such as installing solar panels on roof tops. But care is required in designing such subsidies. The UK solar installation industry went through a rapid boom then bust when the government sharply cut the overly generous subsidy. Subsidies were also set at different size thresholds resulting in sub-optimal impacts. Rather than efficiently using the whole roof for solar panels, owners only covered their roofs up to the 4-kW threshold so that they could receive the maximum subsidy. Similarly, many new micro hydro-electric power stations on rivers have been designed with restricted capacity to maximise subsidies, rather than to optimise the production of renewable electricity.

To aid the further spread of renewable power, regulations should require developers to build solar panels on every new building assuming the local electricity grid can manage it. Rather than covering good agricultural land, solar panels could be retrofitted to most buildings, along railway and motorway verges, in car parks and floating in some reservoirs. Many businesses such as supermarkets, farms and manufacturers already benefit from installing solar panels on their roofs - this should be mandated through regulations, perhaps with access to low cost loans to help pay for the up-front capital cost. IKEA, the Swedish owned furniture chain, installed one million solar panels on its warehouses in 2019. They invested $2.8 billion in renewable energy that year towards their goal to be climate positive by 2030. This shows what companies can do, at scale and cost effectively.

Innovation will continue to bring down the price of generating electricity from renewables by improving efficiency and output. Innovation can also support new forms of electricity generation such as floating offshore wind which will enable turbines to be installed further out to sea, and wave power, although this struggles to be cost effective due to the destructive potential of storms and because the best sites for wave power tend to be remote, far from where the energy is needed. However, Scotland has successfully demonstrated the generation of tidal power from ocean currents in the Pentland Firth. Tidal power has the advantage that its power output is predictable, being primarily dependent on the effect of the moon's orbit. Another emerging technology is kites attached to generators that can access the stronger and more reliable wind at high altitude.

Except for tidal energy, the power generated from renewable sources suffers from uncertainty, particularly in periods of cloudy, dry and calm weather. Another source of electricity is needed which could be supplied by nuclear, giant interconnector lines from other regions or from gas power stations with carbon capture and storage - perhaps not used as baseload but to meet peak demand and when renewable sources are in short supply.

Power companies are already introducing technology to balance the electricity grid to cope with the more irregular generation typical of renewable power. Depending on circumstances, this flexibility can include new transmission and local distribution lines, changing the customers' timing of demand for electricity and creating new flexible energy demands. Alternatively, it is possible to store 'excess' electricity using batteries, compressed air or pumped hydro-electric, or it can be used for other purposes such as to make hydrogen which can be stored for later use. Counter-intuitively, on occasion it can be cost effective to pay companies to stop producing 'free' renewable power. But encouraging energy companies to invest in new storage capacity in a privatised market is challenging. For grid stability, it would be useful to have more pumped storage hydro-electric schemes as they are the largest scale method to store electricity at present, but no privatised energy company is willing to invest because of the level of risk. A new scheme would have a long payback period, whilst future government energy policy and energy costs are uncertain. The common-sense solution is for a more direct approach from government, either to create more long-term policy certainty for private investors or for the government to invest directly.

In future, with the decentralisation of energy, individuals and communities will be able to make choices about where their electricity comes from and how they use electricity. At a local level, house owners may have solar panels on their roof, an electric car charged in the driveway, a battery for storage and smart appliances. Excess electricity generated from the solar panels can heat hot water or be sold to neighbours. A smart meter and smart home hub will enable new interactions between house owners and their local electricity company for mutual benefit. For example, you might need your clothes washed and dishes cleaned by 6pm and allow the electricity company to determine what time your appliances are switched on when you are out at work. And, if you do not need to recharge your car when you get home from work, you could authorise the electricity company to charge it any time before the following morning. If there is a surge in demand, the energy company will be able to withdraw charge from the car battery, your house battery or even switch off your freezer for a short period of time whilst ensuring that the food does not defrost. Conversely, they will charge your car battery overnight when electricity demand is low. So, consumers could give control to a third party within limits but be able to override this if necessary. Life will become a bit more complicated, but in return your power company will pay you or credit your electricity bill.

In summary, all governments should tax electricity production directly proportional to its carbon emissions and phase out and ban coal as quickly as possible. Gas power stations need tight standards to restrict their hours of use, or the carbon emitted needs to be captured and stored underground. A mass roll out of renewable generation is required ranging from solar on every roof, vast new solar farms in deserts to large-scale offshore wind farms; and further innovation is required to best integrate renewable electricity into the grid. The future is a mix of large-scale centralised and decentralised generation from renewable sources with individuals and businesses able to generate their own electricity and agree terms with their energy supplier for mutual benefit. This is an exciting prospect, a vast improvement from relying on dirty coal for our electricity needs.

Chapter 17:
Our Buildings

All my five common-sense principles are relevant to buildings, led by 'embrace efficiency' and 'price carbon pollution'. The focus here is on our homes, although similar principles will apply to offices and commercial properties. Buildings account for nearly half of UK emissions of carbon dioxide. One quarter of this is from mining and sourcing materials used to construct them and the rest to heat and run the homes, offices and other buildings that we live or work in. In addition, mining, extracting sand and gravel and building on previously undeveloped land impacts directly on nature - usually destroying it locally.

Our homes are important to us, and housing is politically important. The UK and Scottish Governments are conscious of fuel poverty, defined where households spend more than 10% of their net income on energy used at home. An obvious way to reduce emissions from buildings is to increase the price of gas and electricity, but this would severely impact those who are least able to pay. In cool, damp countries like the UK the inability to heat homes to a comfortable temperature impacts on health, for example it can cause hypothermia and respiratory difficulties from damp, leading in severe cases to premature death. In hot countries overheating can lead to heatstroke. When officials design housing and energy policy, they should take care to reduce carbon emissions and protect vulnerable people.

Navigating the construction industry is like finding your way through a labyrinth. Typically, a client who wants a new property, or a developer who speculatively wants to build one, specifies the broad type of building they want and its preferred location. A consultant or project manager draws up a design using an architect, surveyor and structural engineer. Finance is negotiated and arranged. A complex procurement process results in the appointment of a contractor followed by sub-contractors including builders, carpenters, plumbers and electricians. Building materials are sourced at the lowest price from numerous manufacturers and suppliers. Despite good intentions at the outset, the tendency is to adopt the lowest cost solutions, often with resultant poor-quality materials, workmanship and quality control. With so many links in the chain, all it needs is for one of these parties to be unfamiliar with green construction techniques or unwilling to cooperate, for the whole project to fall to the lowest common denominator. Consequently, this arrangement is highly unlikely to result in a low carbon and sustainable building. In addition, most new buildings are bespoke, that is a distinctive design needing slightly different materials. Materials are over ordered, just to be on the safe side. There is a significant waste of good quality raw materials, often thrown out direct to landfill without ever serving a useful purpose.

Tighter building regulations are the easiest way to lift the bar of this lowest common denominator, although of course council officials need to check and enforce these. An alternative which might slowly emerge is for a general up-skilling of all the professions and trades involved in the construction industry. If for example, a major property owner, say a supermarket chain, demands high sustainability standards in its buildings, and clearly specifies this throughout the process then this will have a wider impact on all the consultants and suppliers that it deals with.

Another antidote to this construction lottery is to design and assemble building components off-site in a factory; with insulated walls, window-frames, doors and roof trusses cut to exact size in a controlled indoor environment. The factory delivers these pre-assembled components to site then contractors can erect the core of the building and make it watertight within 24 hours. This construction method brings speed, efficiency and quality control and can eliminate most waste of construction materials.

****

Before tackling energy efficiency and heating technology, the first consideration should be the building size and occupancy. A rough pecking order, starting with the best from a carbon perspective, is flats, tenements, terraced, semi-detached, then worst of all, detached homes. Put simply, shared walls and roofs reduce the resources required to build a house and are easier to heat. If two or more people decide to cohabit and move from two single person houses to a shared house, then this will nearly halve the total energy needed for heating. Similarly, we should encourage flat sharing and inter-generational living. This is something we have 'lost' in most high income countries but is normal in many countries where family ties are still strong. From an energy, carbon and perhaps societal perspective it is inefficient and slightly sad to see a single elderly person still living in a large family home with most upstairs rooms unused and an overgrown and neglected garden that they are no longer able to manage.

Stamp duty (Land and Building Transactions Tax, in Scotland) is a rather odd tax as it is paid when people buy houses and so it hits them at the precise time that they are already under huge financial pressure. The tax is also a disincentive to people moving home, acting as a barrier to social mobility, the jobs market and therefore the economy. Abolishing or reforming stamp duty could entice more people to move closer to their work or school and might encourage elderly people to downsize to a smaller, more suitable property. As a first step, the government could abolish this duty for retired people who downsize to a house with fewer bedrooms – perhaps also for people who move house closer to their workplace. Council Tax is another tax to consider. Individuals in the UK can claim a 25% reduction in Council Tax if they live alone. This may seem equitable but may indirectly encourage low occupancy of our building stock.

People buy second homes as an attractive lifestyle choice for those with spare money, as an investment, or both. But many lie empty most of the time, and if they only attract transient weekend visitors, they barely contribute to the local community and economy. The willingness of wealthy individuals to buy second homes both in rural and urban areas can push up house prices across a wider area making it difficult for locals, perhaps on low incomes, to buy or rent a property. In some areas it will worsen the housing crisis and the number of homeless. The prominent environmental campaigner George Monbiot went as far as to say that second homeowners are "selfish" because owning two homes can deprive someone else of a home. This is perhaps less true for newly built second homes, and will vary on a case by case basis, but many local councils already place restrictions on second home ownership; introducing rules about residency before you can buy or levying the full rate of Council Tax on second homes. Raising taxes is one sensible way to discourage second homes and to offset some of the extra environmental impact from those that do own two properties.

A greater environmental impact than second homes is the building of new detached family homes on previously undeveloped land including farmland. Such households are usually completely dependent on private cars to access services. Instead we should continue to regenerate our city centres and undertake a programme to build flats and small houses designed mainly for older people, perhaps with some communal facilities to help combat loneliness and to provide emergency care facilities. These should be built to a high standard to make them attractive for people to choose to move into. We need changes in planning guidance to encourage this shift, and then builders will respond accordingly.

When living or working in buildings we can save energy through a change in attitudes and behaviour. Heat is wasted when windows are open at the same time as the heating or air conditioning is on. This phenomenon is common; in homes, offices, public buildings, halls of residence and hotels; indicating a disconnect between the building occupiers and whoever oversees the heating system. Smart heat and cooling controls can help, but it also requires effective communication with the building occupants to ensure there is no desire or need to override the controls. Also, smart controls must be easy for the householder to operate, as the average person will not be willing or able, to spend the time or effort to understand complicated controls. Another obvious lifestyle change is to wear suitable clothing dependent on the weather. In cold climates, wearing a jumper at home and at work, whilst in hot climates with air conditioning, wearing short sleeved shirts without a tie. When there was a shortage of electricity supply after the Fukushima nuclear incident in Japan, businesses relaxed their office dress code and turned the air conditioning down.

****

We use electricity at home to cook, to run domestic appliances, to operate electrical equipment, to light our rooms and sometimes to heat our water. Meanwhile the UK is dependent on gas to heat our homes and water. Gas became the dominant fuel because of the proximity to cheap natural gas from the North Sea. Now, the UK imports more than half of its gas.

Across Europe, there has been a substantial fall in electricity consumption at home in the last 10 years driven by EU regulations that mandated LED lights and encouraged more efficient domestic appliances. These regulations have achieved a step-change reduction in our energy use. Now they should be widened and tightened to take account of the further innovation potential to reduce energy use. Gas use has fallen too, driven by regulations that require more efficient condensing boilers; and subsidised schemes to install thicker attic insulation, cavity wall insulation and double glazing. Personally, I have installed triple glazed windows in some rooms, but there is no underfloor insulation anywhere in our house, including the modern extension that we commissioned. At that time, underfloor insulation was not a requirement of building regulations even though it would have made our home more comfortable and cost effective to run. Now it would be disruptive and expensive to raise the floorboards to fit insulation. This illustrates that we need tough regulations to force architects, surveyors, construction companies and householders to invest in the best to benefit us all over the long-term.

There is a problem across the construction sector of split responsibilities – those that design and build a property are not normally those who pay the energy bills. The developer therefore has little long-term financial incentive to build to a high energy efficiency standard. Similarly, when a tenant rents a property, on a short-term lease, there is little incentive for either the tenant or the property owner to upgrade to higher energy efficiency standards. Tighter regulations forcing change would benefit everyone.

We need a national campaign (a Green New Deal) of regulation, education and targeted subsidies to compel us to increase the insulation of our homes and offices. Tough regulations need to apply to all our existing building stock: homeowners, private and social rented tenants and commercial properties. They could be applied at certain 'trigger' points such as the sale of a property, an extension being built, or at the end of a tenant's lease, with a back-stop date for all properties to be upgraded.

The costs can be rolled up into a new mortgage if required, whilst business property owners should be able to access loans to help with repayments made from the consequent savings in energy bills. We should target subsidies at specific technologies such as triple glazed windows, as double glazing still leaks heat far more than walls and roofs. In the UK, people pay 20% VAT on refurbishing existing properties, but no tax is due when building new homes. This means that all repairs and improvements, including fitting insulation to existing homes are significantly taxed; whilst tax breaks encourage the building of new homes, often on agricultural land. To reduce further urban sprawl and the inevitable increase in road traffic, it would make more sense to reverse this. Apply VAT to new properties and cut it for energy efficient refurbishments.

For new homes it is much easier to apply tight building efficiency regulations. Germany leads the way with their 'Passivhaus' or passive house movement, building new, and sometimes retrofitting, homes and offices to such a high standard that they do not require a boiler or central heating. Buildings are designed to maximise solar gain, and the heat from human occupants and household appliances is usually enough to keep the temperature comfortable. Developers construct buildings with high quality insulation, triple glazing, good air tightness, heat recovery and extractor fans to circulate air.

We need new building regulations to ensure all new buildings are built to passive house or near zero carbon standards. Housebuilders will complain about the cost, but like catalytic converters in new cars, any extra cost will fall rapidly. Households will enjoy lower heating bills. Every new building should have either solar panels or a green vegetated roof. Roof tiles have already been developed that double up as solar panels which will reduce the cost, whilst regulations could require rainwater to be harvested from roofs to reduce run-off and flood risk, with the water used in gardens or to flush toilets in the home.

Most of the main construction materials used in buildings and new developments are carbon intensive to manufacture. Concrete, steel, bricks and asphalt are all used in vast quantities to construct infrastructure such as buildings, roads and bridges. Concrete alone accounts for around 8% of global carbon emissions - used most extensively in countries that are developing and urbanising quickly. In the UK we have environmental best practice quality standards (such as BREEAM), but these do not always take account of all the impacts from constructing buildings. In Scotland, iconic buildings such as the Scottish Parliament and the Dundee V&A museum are energy efficient in use, but both required large volumes of concrete to construct. It would therefore make sense to introduce regulations, not just for energy efficiency and heating, but also to influence the choice of construction materials. Encouragingly, France recently announced that all new public buildings must be built with at least 50% wood or other organic material.

With a bit of effort and further innovation, the actual construction of homes could be carbon neutral or even carbon positive. Traditionally, bricks were made of local clay, dried in the sun, with a low environmental impact. Bricks and mud are still a common building material in many places, but in countries like Morocco, more permanent concrete structures are rapidly replacing traditional mud brick homes. Innovation could overcome this by developing more durable clay and mud bricks. Architects can also specify building materials that store carbon for decades, the most obvious being timber, straw, hemp or bamboo. Hemp, mixed with lime, makes a building material that is lightweight, breathable and has good insulation properties. Innovation has produced laminated timber; strips of wood glued together to produce a building material that stores carbon and is strong and lighter than steel. It can be used in many buildings, for example, the world's tallest timber framed building, at 18 storeys high, has opened in Norway.   But market forces are not going to reduce our demand for concrete or make the dramatic changes in construction that we need.   Manufacturing cement and concrete is within the EU Emissions Trading Scheme, but the sector has been allocated free allowances to enable it to compete against imports. Taxes need to be raised to make cement more expensive to fully account for the carbon that its production emits, and this would encourage innovation to use other building materials.   The example of France above, demonstrates another way to encourage the shift to materials which store carbon.

****

Providing enough and reliable low carbon heat is a difficult issue particularly in cool damp climates like north-west Europe. The solutions include burning biomass, heat pumps, solar thermal, recovering waste heat or burning 'green' gas whether this is biomethane or hydrogen. We can apply these solutions to individual properties or to multiple properties through gas pipes or district heat networks. Following the rise in oil prices in the 1970s, Denmark, which did not have access to its own natural gas, invested in district heat networks where underground insulated pipes transfer hot water generated from a central source to homes and businesses. The Danes have now connected two-thirds of all houses to district heat, rising to 98% in Copenhagen.

Burning biomass to produce heat is like burning coal but has the advantage that it is near carbon neutral assuming that the trees or crops are replaced as fast as they are harvested. But away from rural areas, and forested countries like Sweden, there are limited supplies of biomass. Biomass for heating is best restricted to rural areas; near the source of wood and to prevent air pollution in cities.

Heat pumps work at maximum effectiveness if the source of water is warm and the water circulated to heat homes is at a lower temperature than what is normal with natural gas. So, they work better in buildings that are already well insulated, and with underfloor heating or large radiators. In addition, a tank is required to store hot water, and in very cold weather householders may need to switch on the heat pump earlier and for longer to reach the desired temperature. Consumers will therefore need to be persuaded that this technology is suitable for them.

With heat pumps there is no combustion within your property, so there is no risk of explosion and no emissions of nitrous oxide. But heat pumps use a lot of electricity, particularly during very cold weather (or very hot weather if used for air conditioning) and, if whole streets were to fit them, this would place a huge strain on any electricity network. Some solutions to this were considered in the previous chapter on generating electricity.

District heat pipes are simply a method of transferring heat into buildings. In the UK, most district heat networks burn natural gas to provide the heat. A few years ago, this was a lower carbon option than heating homes from electricity, but not any longer as the electric grid has been substantially decarbonised. Burning gas for district heat is not a low carbon solution. Therefore, district heat needs to be supplied from renewable sources such as solar thermal, biomass and heat pumps, or by capturing the waste heat from industry. In 2010, Scottish company, Star Refrigeration installed a water sourced heat pump to heat a district heat network in the Norwegian town of Drammen. Each unit of electricity produces three units of heat, which is distributed in pipes at 90°C to heat most of the city. Scotland is now installing its first large-scale water sourced heat pump as part of an urban regeneration project at Queens Quay in Clydebank. Most towns and cities are located on rivers or by the sea so the potential for large-scale water sourced heat pumps is significant. Other sources of pre-warmed water include abandoned mine works, sewage wastewater pipes and wastewater from industrial processes.

It is easy to install district heat in new housing and commercial estates, but it may not make commercial sense. New buildings with good insulation have a low demand for heat which heat pumps to individual properties can provide cost effectively. The bigger role for district heat is to retrofit it to existing properties in dense urban areas, albeit this will necessitate a capital cost, and disruption to roads and pavements to lay the network and local pipes to every property. Councils or communities will need to persuade local people and businesses that this short-term disruption will be to their long-term benefit.

For most consumers, continuing to use the existing gas network, and burning 'green' gas, would be the least disruptive change. Organic material can be input to an anaerobic digestion plant to produce biomethane to inject into the gas grid. Although this might seem like a good solution, there is a limit to the available supply of organic material such as sewage, waste food or crops. An alternative is to transmit hydrogen through the existing gas network. It is technically possible to add up to 20% hydrogen into our existing gas grid, but we would need 100% to replace all natural gas. This will require the gas utility to change all our boilers and gas cookers to be compatible. Hydrogen is produced from steam methane reforming of natural gas, but this is energy intensive with a 'loss' of energy on conversion from methane to hydrogen. And, to be carbon neutral, we would need to capture, transport and safely store the carbon dioxide produced as a by-product – all require further energy and cost making this an unattractive prospect. A better solution is to make hydrogen from electrolysis powered by renewable electricity. This is currently an even more expensive process and also suffers from high energy conversion losses. Only if the price of electrolysers and renewable energy, such as solar or offshore wind, continue to tumble will it make sense for hydrogen to become part of the solution to our heating issues. Its great advantage is that you can store hydrogen to manage peaks in the demand for heat.

There are several potential ways to provide low carbon heat, but none are easy. We will end up with a patchwork of different solutions in different geographic areas dependent on housing stock and local geography.

How do we drive these changes in our heating and wean ourselves off natural gas? Firstly, gas is too cheap; the price we pay does not factor in its impact on air pollution (nitrous oxide) and climate change. Secondly, changing the balance between the price of gas and electricity would help - the so called 'spark spread'. This spread is artificially high in the UK because of government taxes and levies which favour cheap gas and increase the cost of electricity. In the UK, a unit of electricity costs around four times that of gas although this varies depending on commercial agreements, access to local generation and the time of day. Given that electrically driven heat pumps are more expensive to buy, and then produce two to three units of heat for each unit of electricity, they will always struggle to compete against subsidised gas. In effect government policy inadvertently discriminates against heat pumps even although they are the most efficient low carbon heat system. In contrast, Norway, with its cheap hydroelectricity, has a lower spark spread and they are deploying far more heat pumps.

We should therefore tax gas to take account of its impact on the climate. There are several ways to do this. One is to levy 20% VAT on consumers, another less controversial is to transfer some of the environmental taxes currently levied on electricity across to gas. However, there are limits as putting up the price of gas to a sufficiently high price to force us to stop using it would probably be political suicide, especially as it would hit poorer households and vulnerable businesses the most.

The capital cost of moving to a low carbon heat system will be prohibitive for many, and the cost will be even more expensive if individual properties change one by one over several years. It would be far more efficient for local councils to zone areas into recommended low carbon heat types and to encourage streets or districts to convert together. For example, there could be six main categories: industry, low density new homes, low density existing homes, new city centre developments, retrofit existing city centre buildings, and retrofit tenements, flats and terraces. To ensure the economies of scale required to make this change cost effective, councils should give commercial properties and public sector buildings a limited time to transfer to the new heating system, enabling energy suppliers to offer heat pumps or district heat to all the residents of a street or district.

Introducing 'heat as a service' business models might help to smooth the conversion to new low carbon heating solutions. The energy supplier would install insulation and a new heating system, maintain it and provide heat to the customer at an agreed temperature at a fixed cost over a lengthy contract period. This provides an incentive to install a quality product; the customer avoids any capital expenditure, has the comfort and security of a fixed inclusive monthly price and does not need to worry about repair bills.

Improving the woeful energy efficiency of the UK's buildings and moving to low carbon heat systems is a national initiative that may take two or three decades to complete. It will create thousands of new jobs, for example, colleges retraining gas installers to install heat pumps. The jobs will arise in manufacturing, design, installing insulation, boilers, pipes, constructing new energy centres and maintenance. We also need to generate more electricity, strengthen the grid and develop new customer contract options. The benefits extend beyond jobs; we will enjoy better insulated warmer homes, better health, safer heating systems, stable energy prices and will no longer be dependent on energy imports.

****

Living in a country with a cool, temperate climate, I have focused on our heating needs, but cooling is more important in many countries. Global demand for cooling and refrigeration is growing rapidly. It may double or even triple by 2050. India alone had two million air conditioners in 2006, with 200 million predicted by 2030. Urbanisation, increased affluence and adapting to a warmer climate drive this increase in demand. Some natural cooling can be provided through good architectural design; avoiding large south-facing windows, use of shade, appropriate building materials and solar chimneys in the centre of buildings that naturally draw warm air up and out of the building. Trees and green roofs, or alternatively painting roofs white to reflect solar radiation will all reduce heat in buildings. Electric air conditioning units, or less commonly, efficient heat pumps (working in reverse) often supplement these. Paris is developing a district cooling network, cooled by water from the River Seine.

The great natural advantage of cooling is that in hot countries the demand is closely correlated to the potential supply of solar electricity. Heating by contrast, is most required in the winter months in high latitudes with little solar radiation. The electricity required for cooling can therefore be provided from solar panels especially if more energy efficient heat pumps are used for cooling rather than direct electric fans. However, the other issue is the refrigerants used in some cooling and refrigeration. The Montreal Protocol has banned HCFCs to tackle the destruction of the ozone layer, but often they have been replaced by hydrofluorocarbons (HFCs) which, whilst not affecting the ozone layer are powerful contributors to global warming. Governments need to accelerate the proposed phase out of all these gases, currently not due until 2036.

In summary, for the UK, we need strong action by government to encourage and persuade house owners to change from gas to low carbon heating systems. We need a long-term plan and to take investment decisions with long payback periods. In fact, more than most other areas of our lives, government led regulations are essential to overcome the inertia in our choice of heating system. Getting the basic taxes right for gas and electricity is an essential first step, followed by regulations, planning rules, subsidies and new finance models. Moreover, we should safeguard the most vulnerable in society, and adopt an education campaign to persuade the public and business of the necessity for this change. Council or community led projects might also help to bring the public on board. Society will look back in a few decades and wonder why we ever allowed developers to build houses that needed a heating source to keep us warm and comfortable, and why we allowed energy companies to send a combustible gas into our homes that caused local air pollution and emitted gases that warmed our climate. Converting to heat sourced locally will create new jobs and avoid relying on imported natural gas. A future where we all live in warm, comfortable homes is better for us all.

Chapter 18:
Surface Transport

Of my five common-sense principles, clearly 'travel wisely' is important, followed by 'price carbon pollution' and 'embrace efficiency'. The focus here is on our private travel by car, train, bus and bike across land although I also consider logistics – ships, lorries and vans used by business to deliver goods and services. A later chapter on tourism and leisure considers ferries and leisure trips on boats.

The global demand for travel has been increasing for decades, with flying growing the fastest, followed by car travel. Population is increasing, but in addition the distance we travel increases as we get wealthier, with no sign of this abating. The demand for international travel is a recent phenomenon, driven by the globalisation of business supply chains and our ever-growing desire for long-distance leisure travel. Travel within countries is increasing. For many years, people in high income countries have been moving out of city centres into ever expanding suburbs and rural locations with no local services. In low income countries people flock to cities, often moving from a lifestyle with little motorised transport to city commuting and occasional trips back to see family in the countryside. High income countries have also become reliant on out of town shopping centres, business parks and industrial estates which fuel the need to travel, normally by private car as they are usually located on the outskirts of towns near ring roads. Fundamentally, land use planning decisions are at the heart of our need for local travel. Mistakes made stay with us for decades.

In dense urban areas there is a critical mass of people to enable public transport to run frequently and efficiently.

There is limited space to park cars, so people in cities like New York and Hong Kong have a much lower travel carbon footprint than those living in towns and the countryside. In high income countries, visionary planners have made several attempts to design car-free towns. Zermatt in Switzerland took the decision to restrict cars from entering the town, cars are excluded from central Florence, and Avoriaz in the French Alps is a purpose-built ski centre designed to be car free. These examples have proved to be successful as part of the marketing of these towns, but they are artificial as they all rely on tourism. An example of wider relevance is Oslo, the capital of Norway, which is car-free in its centre. Access is limited to the disabled, freight deliveries and the emergency services.

There are many ways to discourage private cars and encourage public transport, walking and cycling. These measures are push and pull or carrot and stick, and only need to be enough to tip car owners into choosing not to use their car. These include making streets for pedestrians only, charging cars to enter certain areas, park and ride facilities, low speed limits, and investing in good public transport, bus lanes and cycling infrastructure. Congestion charging has reduced traffic in central London and encouraged more cyclists. Restrictions on the availability of car parking might not be popular with car drivers but can also be effective. Planners and councils can restrict the number of car park spaces in new housing developments, introduce resident only permits, plant trees in suitable streets, or implement work-place car park charges. All these changes will help nudge people out of their cars in cities, but there is a risk that some could backfire. Park and ride schemes for example, might solve city centre congestion, but may encourage more people to live out of towns and therefore travel further. We need a package of measures including restrictions on new homes in the countryside.

Owning a private car, a tonne of steel and electronics, only used around 5% of the time, is not very efficient – and 35% of households in the UK have access to two or more cars. Cars have given us 'freedom' to move around for work, leisure and to visit friends and family. But they have also enabled us to live further away from work, education, family and local leisure activities requiring longer travel and commute which leads to stress, congestion and air pollution. Many of us depend on our cars as we live or work where there is no access to reliable public transport. But even in the UK one quarter of households do not have access to a car either through choice or because they cannot afford one.

Cars and vans with traditional fossil fuel-based engines emit carbon dioxide, particulates and nitrous oxides from their exhausts causing local air pollution which causes some people breathing difficulties, ill health and even deaths. It can adversely impact children's brains as they develop. Cars, regardless of fuel source, require huge amounts of land for roads, garages and car parking, whilst street parking and congestion prevent children from playing outside their homes. Congestion slows down bus journeys. Rural motorways make it difficult and dangerous for wildlife to move around, and urban motorways create an artificial barrier which prevents pedestrian access between neighbouring communities.

A few decades ago, owning a private car would be considered a luxury. Even today it is quite an extravagance, but we perceive it will improve our quality of life. However, cars place financial burdens on society and adversely impact the lives of others through noise, air pollution and accidents. They make our cities less pleasant places for others to live in - think of an elderly person trying to cross a busy road. In addition, congestion can waste time for drivers along with its resultant frustration and stress. There is a vicious circle in play here. As streets become more congested and dangerous to cross, parents feel obliged to give their children a lift to school. As air pollution worsens in cities, people attempt to filter the air with their air conditioning, and more move out to the suburbs with a long commute back to the city.

Electric cars will certainly improve local air quality and reduce carbon emissions. But they will not solve all problems and will create some new ones. Their heavy batteries require metals such as cobalt, lithium and nickel. Mining is environmentally destructive and is concentrated in a small number of countries. In the Atacama Desert in Chile, miners inject precious scarce water into underground pools to force the saline water to the surface to concentrate the lithium through evaporation. Some of this water is produced from energy intensive desalination then pumped up mountains to where it is needed. Chile has more than half all known reserves of lithium, whilst other metals come from countries prone to conflict and corruption such as the Congo. Our society is in danger of becoming as reliant on these countries as we currently are on oil from the Middle East. Electric cars are heavier than fossil fuel powered cars and therefore there is more wear and tear of tyres. Tyres gradually wear down releasing micro-plastic into the environment which ends up in our soils, rivers and oceans.

Many futurists dream of a future with self-driving electric cars and suggest that this will solve these environmental problems. If we shared such cars and used them constantly then we would need fewer cars. There would be less congestion, time wasted searching for a vacant car park space and less land taken up by parking spaces. The software could select the most fuel-efficient route and speed. Computer controlled cars with automatic brakes could drive close together reducing wind resistance. This would enable more cars to be on a road at the same time or the authorities could reduce the road width to make room for cycle lanes or wider pavements. But access to self-driving cars is likely to result in an increase in total distance travelled and may increase congestion. For the first time, everyone would have access to a car, people could commute much further and work or sleep during the time spent travelling. In any case, people like to own status goods like cars and might still choose to own a self-driving private car rather than hire one when they need it.

Trains, buses and coaches are a more efficient mode of transport than cars because of the number of passengers they can carry. Trains are best because they are narrow with low air resistance in proportion to the number of passengers carried and there is a low rolling resistance, or friction, between their wheels and the rails. However, air resistance increases at higher speeds and this has a significant impact on fuel economy for trains and road vehicles. The Energy Savings Trust estimate that the fuel consumption of a car travelling at 60mph (97kmh) is 15% lower than one travelling at the legal speed limit on UK motorways of 70mph (113kmh). Across the UK some railway lines are being electrified, but most buses, coaches and lorries still burn diesel. The air pollution from these vehicles is being dramatically curtailed through the EU regulation for manufacturers to fit 'Euro 6' engines for all new vehicles, but unfortunately whilst this technology tackles air pollution it does not decrease carbon emissions.

Cycling is the ultimate in low carbon travel, but to encourage uptake, it needs to become more mainstream, to be safer, and most importantly to be part of a wider package of land-use planning so that we live within a short distance of schools, work and access to services. Separate cycle lanes, to keep bikes and cars apart, are better if there is space available. In Copenhagen, whose climate is not much different from the UK, around 60% of people choose to cycle to work, school or university. This shows what a compact city with good cycling infrastructure and a supportive culture can do.

****

To reduce the carbon impact of our travel, our priority should be to reduce our demand or need to travel and to encourage us to walk and cycle. For longer distances the most sustainable solutions are mass public transport, shared vehicles then single-occupied private cars. We also need to move to low carbon methods of propulsion such as electric or hydrogen fuel cell.

In the long-term, we need pro-active land-use planning to reduce our need to travel and our dependence on private cars. Compact towns and cities are best – a key concept is that everyone should live within 15 minutes walk, or a short cycle, of major services such as shops, schools, parks and good public transport links. We could make huge progress towards achieving this, but only with much stronger and proactive planning rules instead of planners reacting to proposals from developers. Planners should discourage, or ban, new homes in the countryside that inevitably create a dependency on cars to access services. Meanwhile, our towns and cities can become more attractive with green streets – trees planted in pavements and along the sides of streets, fewer car parking spaces and better provision of public transport.

The internet influences our choices and increases our options for work, education, shopping and leisure. It can be a powerful force to reduce travel; enabling home working, shopping from home and virtual leisure experiences but there are counter effects that will reduce any carbon savings. Home shopping has led to a huge increase in delivery vans; the internet makes us more aware of overseas destinations which can lead to a desire for more travel; and the ability to work from home can encourage us to make new work choices. For example, we might accept a long distance commute two or three days per week if we can work from home on the other days. Overall, it is the choices that we make from our use of the information and capabilities of the internet that will determine whether it increases or reduces our travel.

An issue often ignored is that parental choice of schools leads to more travel and peak time congestion, whether this is for private education, religious reasons or to access specialist facilities. Encouraging all children to attend their local school would cut the need for parents to transport their children around cities, and would enable most children to walk, cycle or share a bus to their local school. This may be controversial with some, but we need to think afresh. Reducing choice may have the benefit of parents increasing the pressure on local schools to improve their performance. To encourage cycling, schools should provide facilities, teach cycle safety and promote cycling to parents and children.

Encouraging more people onto public transport would free up road and car parking space. But, for those who already own a private car, it is often cheaper to use it than to take public transport. We need to shift this balance, by making the use of private cars more expensive, or less attractive in comparison to public transport. The obvious options are to increase fuel duty and car park charges; whilst providing high quality, subsidised public transport. Luxembourg now provides free public transport to all its residents. To encourage a shift out of cars, and not to lose public support, it is important to give people time to adapt and to make these changes in tandem. For example, councils should use the money from car park charges or tolls to provide better public transport.

Many people love their cars, some dislike public transport. It will be difficult, and unnecessary, to persuade all to abandon car travel. What we can do is reduce the need to travel by car and to reduce the impact of that travel. We should ban or heavily tax high carbon emitting cars so that people no longer think of them as a mainstream purchase. The government should design company car tax, car purchase tax and annual car tax (vehicle excise duty) to encourage people to choose fuel efficient, low emission cars.

Having to pay a high cost to drive each kilometre nudges people towards public transport and car sharing – something we have become lazy about. In the UK, fuel is already relatively highly taxed, attracting fuel duty and VAT, however the government has frozen these tax rates for the last ten years, and have in effect reduced them due to inflation. As a result, the total number of kilometres driven in the UK has nudged back up following several years of slight decline when the 'fuel tax escalator' was in force. It may not be popular with many car drivers and hauliers, but a steadily increasing fuel duty would be better for our society.

The transition to electric cars will soon cause a problem to the government and society as the income raised from fuel duty will fall steadily. It will be difficult to raise the tax on general electricity as this will hit people on low incomes. One option is, like income tax, to levy an escalating rate of tax on domestic electricity consumption. Wealthy people who use a lot of electricity and own two electric cars would pay a higher rate. Another option worth exploring is whether it is practical to restrict any tax increase just to electricity supplied to specialist car chargers. If this is not possible, and there is no tax based on distance travelled, then it is likely that we will all choose to drive more, leading to further congestion. So, it is in all our interests that we design a replacement for fuel duty, accepting toll booths as in France or preferably, although more complex, a tax per kilometre travelled.

How can the government encourage us to shift to driving electric cars? Banning diesel engines from city centres to combat air pollution will encourage us to think more about buying an electric car, even those of us who only visit city centres occasionally. We may continue to need direct subsidies to lower the purchase cost of an electric car or to help install charge points at home. Batteries make up a substantial proportion of the purchase cost and these are becoming more effective and cheaper, and car production costs will continue to fall as economies of scale kick-in at car manufacturing plants. So, we can phase out subsidies over time. Company cars account for half of new sales in the UK and after three or four years they become available to the public as second-hand cars. So, by manipulating company car tax to favour electric cars, the UK Government can introduce a rapid change across the car fleet. To ensure a complete shift to electric cars, governments should ban the sale of new petrol and diesel vehicles - an early date for pure petrol and diesel and a later one for hybrid cars and vans.

Governments and councils also need to provide the right framework for an extensive electric charging network. Most early adopters of electric cars charge at home, or at work, but others rely on street parking. Innovation will help; lampposts can be rewired to provide energy to charge a car overnight on street parking, charge points could be attached to street bollards and trials are underway to charge taxis and buses without cables from plates embedded within the tarmac of taxi ranks and bus stops.

We should encourage cycling for health and sustainability reasons, but it is not going to replace public transport and private cars. Here price is not really the problem. We need good cycle infrastructure such as cycle lanes, clear road markings, bike racks, secure storage areas and changing facilities at work. Also, as the price falls, the use of electric bikes will soon become common.

****

Surface transport is not just about our personal travel for leisure or for work. It is also the logistics of companies moving goods between ports, or manufacturing sites to warehouses and then to shops, and includes the recent growth of vans delivering parcels to our homes.

Like personal travel, the priority to reduce carbon emissions is to reduce demand.   Not that companies are likely to deliberately sell less value.   Instead they can consider new business models; local production perhaps using 3D printing, electronic based services rather than physical products, and concentrated products that are less bulky to transport. These can be combined with optimal warehouse locations and delivery routes and avoiding half empty deliveries and empty return journeys.

Other fuel saving initiatives include modal shift, for example, a shift from lorry to train.   Fuel efficient driver training, speed limiters, regular vehicle maintenance and correct tyre pressures will all help, as will buying fuel efficient vehicles with a streamlined shape, low rolling resistance tyres and light-weight chassis.   High fuel prices would encourage and incentivise all these initiatives.   Once these initiatives have been exhausted, companies can consider switching to low carbon fuels such as biodiesel, electric or hydrogen.   Many hauliers operate between limited numbers of sites where it will be possible to set up electric charge points or hydrogen fuel stations.   Given that these alternative fuelled vehicles have a high up-front cost but low running costs, they are likely to be cost effective for delivery vehicles that are driven a long distance.   The only missing piece in the puzzle is tax, particularly fuel duty which the government has frozen for ten years. A steady and predictable increase in fuel duty should be enough to kick-start a wholesale transformation to electric haulage vehicles and vans. Vehicles in the UK also pay vehicle excise duty with different rates for cars, vans and lorries. Policy has changed frequently over the years, but at present, for new cars the duty is correlated to the vehicles carbon dioxide emissions, but for vans it is a standard rate of £250 per year.   It is fundamental that all these taxes and duties are correlated to carbon emissions with high emitting vehicles being heavily penalised.

There is currently a Betamax versus VHS videotape style competition between proponents of electric and hydrogen fuelled vehicles. Like VHS which became triumphant, it looks like battery electric systems are winning, at least for the private car market. Compared with hydrogen they more directly and efficiently convert energy into useful power. Hydrogen requires an additional energy intensive process to produce. A fuel cell converts the chemical energy of a fuel, such as hydrogen, reacting with oxygen into electricity without combustion. Proponents state that hydrogen fuelled vehicles can be lighter than electric ones, with fast refuelling and a long range. Both hydrogen and electric vehicles are continuing to improve their power density and have the advantage of requiring fewer moving parts than cars powered by fossil fuels. However, a transformational shift to electric vehicles may have expensive knock-on consequences on the need to strengthen the electricity grid, whilst arguments against hydrogen are that it will require a new network of fuel stations and it is less energy efficient as the gas must be created then compressed for storage. From a carbon perspective it is important that any hydrogen we use is created from low carbon power sources.

Biofuels are another option to power vehicles. Biofuel from crops is problematic for all modes of transport because of its relative land-use inefficiency. An average solar panel captures 16% of solar radiation to convert to useful energy, but photosynthesis for crops only achieves 2%. Biofuel made from food waste and sewage sludge could make a useful contribution, but there are limited supplies.

It may be that hydrogen fuel cell vehicles will have the edge in heavy duty transport that carries heavy payloads or requires fast charging. This includes lorries, long-distance coaches, large plant and machinery and tractors. Hydrogen is being trialled for trains running on lines that have not been electrified, and hydrogen or methanol may be a useful fuel for the shipping industry. Which fuel dominates will not just depend on cost, effectiveness and the rate of technological improvement. Government policy and support, strategic decisions made by transport manufacturers and what the consumer is willing to buy will also be an influence.

****

International shipping such as container ships and oil tankers are a cheap and relatively carbon efficient method to transport goods around the world. Sailing slowly is fuel efficient as fuel consumption increases at higher speeds. But most ships burn 'bunker fuel', the lowest quality heavy fuel oil, high in sulphur that governments do not permit cars, buses or lorries to use. Shipping emits around 3% of global carbon dioxide; but 11% of sulphur dioxide and 23% of nitrous oxide. It also creates ozone and millions of tiny particulates that can become trapped in our lungs. Ironically, the air pollution from ships (and other sources) may cool regional climates, as the aerosols reflect some incoming sunlight both directly and indirectly by providing nuclei for raindrops to form into clouds.

Shipping has many other environmental impacts including inadvertently carrying invasive species across oceans, underwater acoustics which can harm marine mammals, paints used to prevent marine life attaching to hulls can be toxic, and numerous deliberate and accidental oil spills. The difficulty of solving the environmental impact of ships is compounded by the international nature of the industry, with a tendency for ships to register under jurisdictions with low standards, whilst fuel can be bought at whatever port sells it for the lowest price.

It is outside the scope of this book to outline the international agreements and myriad solutions needed to clean up international shipping, but many of the same principles apply as elsewhere. We need fuel taxes to account for the carbon and air pollution that ships cause, and regulations can force the electrification of power used by ships in port. Further innovation can introduce new means of propulsion such as biofuel, liquid petroleum gas, electric, ammonia or hydrogen. Electricity stored in batteries for short journeys such as ferries, or hydrogen from renewable energy look likely to be the best contenders. Fuel consumption can be reduced by installing 'sails' controlled by software to harness the power of the wind.

In summary, we could reduce our need to travel through re-organising business and logistics and re-considering where we choose to live. This might also be an opportunity to improve our quality of life. Private cars have their place, particularly for rural travel, but they blight our cities so councils should encourage public transport. A wholesale shift to the electrification of cars, and buses, lorries and trains, will benefit us all, particularly through breathing cleaner air. But society needs to take care in this transition, as there is the potential for cheaper running costs to lead to an increase in congestion. If we could accept restrictions on private use of cars, then this would provide an opportunity to radically redesign our towns or cities to be more compact and be better, healthier places to live - car parks could be turned into green parks. And, in future cyclists may no longer need to sit in congestion directly behind vehicles belching damaging exhaust fumes into their lungs.

Chapter 19:
Flying

This is a difficult chapter, writing as someone who enjoys overseas travel and has made useful business connections with people at international events. I certainly do not enjoy the experience of airport security and the endless waiting and queues, but I do enjoy and learn from the destinations and experiences that flights can enable. Tackling the emissions from flying is a problem, unlike most carbon issues, with no current or near future technology that we can deploy any time soon. Unlike shipping, where simply travelling at lower speed will reduce carbon emissions, most passenger jets already fly near the optimal speed for fuel efficiency. For jet engines, around 800kmph is an optimal balance between aerodynamic drag and the extra energy required to overcome gravity for a longer time if the speed was slower. Of the five common sense principles, clearly 'travel wisely' is important, whilst if we 'priced carbon pollution' this would take us a step in the right direction to tackle the emissions from aviation.

Flights account for 2% of global greenhouse gas emissions and, as explained earlier, we should double this to account for aviation's other climate impacts. Until Covid-19, the unstoppable growth in demand and the lack of short-term technological solutions to its climate impact, meant that we needed to reconsider our recently adopted addiction to flying. Now, it looks like the aviation sector; particularly investment in new planes and long-haul flights, will be affected for a number of years by Covid-19 restrictions and loss of business and consumer confidence.

The number of global passengers doubled between 2003 and 2019. This massive growth is recent. It began in high income countries, and more recently spread to the fast-developing economies of India and China. Around 12% of these flights are for business travellers. Business travellers are typically twice as profitable to airlines as leisure passengers and there is a good fit for airlines between business flights during the week and leisure flights at weekends. Freight accounts for around one fifth of the impacts from aviation, although this is difficult to calculate as dedicated cargo planes and the holds of passenger planes are both used to carry freight.

Given its size and wealth, the USA emits the most carbon from aviation. But citizens of Australia and the UK emit more per person than any other country. In the former case because of their relative isolation from the rest of the world, and in the latter because the UK is a relatively wealthy island, with poor high-speed rail links and a temperate, but unpredictable climate, all of which encourage people to choose to fly, often overseas on holiday. An efficient aviation sector that offers extremely cheap flights supports this demand. The result is that the average Briton emits ten times the emissions from aviation than a person from China, although emissions there are growing rapidly in line with the aspirations of their ever-growing middle class. Yet, affluent people still dominate aviation and its impacts. In the UK, 10% of the population make 60% of flights and over half do not fly in a typical year. Imposing taxes on aviation will therefore affect the wealthy more than the less well off.

However, as people become more affluent and as ticket prices fall, our expectations on flying continue to change. As a child, it was a luxury to fly to Spain, now some make multiple flights each year including long-haul holidays. People fly overseas to watch sports events, to see a 90-minute game of football, and others commute by plane on a weekly basis.

It is easy to criticise frivolous flights, for example, weekend stag and hen party trips to Europe. Some welcome such trips, but others feel that they are expensive, time consuming and unnecessary. Friends persuade them to attend through the subtle application of peer pressure. It would not take much to reverse this peer pressure so that people frown upon 'frivolous' flights. Some parents are already finding that their teenage children do not want to fly long-haul for a holiday.

Several companies offer day trips flying from the UK to Lapland to take families with children on a pre-Christmas treat to see snow and Santa Claus. This is affordable because there is no price on the carbon pollution of aviation fuel. Could you justify this to your child in the future when they ask what you did to tackle the 'climate emergency'? Will your child feel unloved, or that they have missed something if you did not take them on this trip?

Honest thinking about the impact of your flights is morally bewildering and an ethical and philosophical challenge. The carbon emitted is invisible and its effects are distant in time and space. Clearly behaviour and attitudes around flying are complicated and vary between individuals. Even if people were fully aware of the environmental damage arising from their flights, most would still find a justification to take a leisure flight in their own minds. This could be one or several of the following:

- I did not really understand the environmental harm it caused, or the flight is going anyway so it will make no additional difference.
- I have worked hard, have paid my taxes and deserve a holiday (or my children deserve a good holiday).
- It is cheaper, or quicker, than the train; or it is the only practical way to reach an island or isolated location.
- My individual boycott will not make a difference; or everybody else is doing it, why shouldn't I?

- The trip is to visit family or friends, is for charitable purposes, will bring money into fragile economies or has cultural or educational value.
- It is the role of governments to regulate, not for individuals to choose.
- I have paid to offset my flight, so I have a clear conscience.

But flying clearly does contribute to climate change and airlines quickly respond to any increase in demand by putting on new services. It might seem that individuals cannot make any difference, but everything we do influences those around us. If you fly for a weekend break, then your friends may feel that they should do the same. If you choose to holiday at home or travel long-distance by train, again, this will influence those around you. The following chapter on tourism considers the argument around charitable, educational, and cultural trips; and a later chapter considers the exaggerated claims by airlines that promote carbon offsets.

The problem is that most of us could argue our case using one or more from the above list. But as a society we do need to reduce our addiction to flying. Flying should become a 'treat' again, rather than routine.

Societal expectations have changed fast, but they can also reverse. If you tell your friends that you are going on a weekend break to Prague or to Thailand for your summer holiday, do they ask lots of questions and admire you for it, or do they politely suggest that it is not compatible with climate change? There has been a culture of encouraging each other to fly more, travel further and fly to new exotic locations. This peer pressure could reverse as evidenced by Sweden in 2019. The number of flights taken fell by 4%, the first fall in decades, and all down to the actions of one young person speaking out. I admire Greta Thunberg, the young school striker and climate change activist who travelled to the United Nations in New

York on a sailing yacht. She then planned to travel to Chile overland to speak at a climate change conference. The UN unexpectedly switched the conference to Spain, and Greta managed to hitch a lift on another yacht back to Europe. Few of us have her resolve and determination.

Behaviour changes and peer pressure can certainly impact on our demand for flights, but it will be difficult to dramatically reduce demand just through voluntary behaviour change. Video conferencing can reduce the need for some business flights, but makes it easier to trade overseas, which may result at some point in travel which a lot of people see as a perk of their job. And, more people now live and work abroad, often laying down roots, with the result that families understandably travel long distances to visit their loved ones. Peer pressure will have its greatest impact on what are arguably more frivolous flights, for example, weekend breaks.

****

It might seem obvious that we should not subsidise aviation and that we should tax it to take account of its carbon and other adverse environmental impacts. But an underlying issue is that governments cannot tax kerosene under the 1944 Chicago Convention. Like shipping, given the global nature of aviation, even if one country did attempt to tax kerosene, airlines would simply refuel at cheaper, untaxed locations.

The main manufacturers, Boeing and Airbus, have already designed aeroplanes to be relatively streamlined but their basic shape has not changed in decades. Laminar flow control can reduce drag around the wings and blended wing technology could merge the fuselage and the wings into one smooth structure. This shape would allow the entire aircraft to generate lift and minimise drag, decrease fuel consumption and enable them to carry a larger payload. But, despite these significant potential improvements to reduce fuel consumption, the culture of safety has led to risk-averse incremental change such as using lighter composite materials. Neither governments nor airline manufacturers have the appetite to implement radical change to the standard design of passenger aeroplane due to the cost, long payback and risk.

We should increase tax on aviation and then steadily increase it further to take account of the carbon pollution and other climate inducing impacts of aviation. This will focus industry minds to innovate and invest in modern fuel-efficient technology. Such taxes should be directly proportional to the carbon emitted per passenger, so it would be higher to travel first class or long distance, and lower for fuel efficient planes. Governments should use some of the tax receipts to fund global innovation competitions and to pilot and prove new technology.

But currently taxes are low. Even worse, governments offer a wide range of subsidies to manufacturers, infrastructure providers and airlines; whilst they frequently own or prop up failing airlines. Governments justify this on national security grounds, to support jobs, or to support regional economic development and business growth. Subsidies include grants, equity, loans, monopoly rights, and no or reduced taxes. In the UK alone, fuel tax, VAT exemptions and duty-free concessions amount to a subsidy of over £10 billion per year.

The Scottish Government provides additional subsidies, with some justification, directed to 'lifeline' services run by Loganair to the remote Highlands and Islands to support access to healthcare, education, business and tourism. A taxpayer funded agency owns 11 remote airports; flights to the Highlands are exempt from Air Departure Tax; people who live in remote areas are eligible for direct flight subsidies, and until recently grants were offered to develop new international aviation routes.

As an alternative to fuel taxes, the UK government introduced Air Passenger Duty (Air Departure Tax in Scotland). It ranges from £13 per ticket for a flight under 2,000 miles in the lowest class, to £172 for a flight over 2,000 miles in other classes. It would be better to reform this tax so that there is a much clearer and more direct link between the tax and the actual carbon emitted.

Meanwhile, the EU added aviation to their Emissions Trading Scheme in 2012, but free allowances are handed out to airlines, and flights going outside the EU are exempt. There is enormous scope to tighten these rules, although this would be much easier, and better with international cooperation.

Taxes alone are not likely to be enough to reduce our addiction to flying. If governments levied taxes at a sufficient level to make a big impact, this would be extremely unpopular with business and individuals who regularly fly and might be politically and economically unfeasible due to the downturn caused by Covid-19. Another option is to introduce a cap on the availability of flights. Rationing is simply placing a restriction on supply; many would argue that the National Health Service already rations the availability of medical treatment in the UK, and any cultural or sports event with a limited number of tickets on sale is a form of rationing that we all accept. A clearer example is that during and after the Second World War, the government rationed food in the UK because food supply was less than demand and the government wanted to avoid price rises. Rationing had an unexpected beneficial impact on our diets, forcing us to eat less and more healthily.

Rationing of flights need not be as extreme. A cap on the carbon emitted from global aviation is an indirect form of rationing. The International Civil Aviation Organisation is already implementing this principle. From 2021 onwards, the plan is to stabilise, or cap, global carbon dioxide emissions from international aviation at 2020 levels (subsequently changed to 2019 to take account of the Covid-19 decline in aviation). Airlines that emit more will need to offset, for example, by planting trees or investing in renewable energy in other sectors of the economy. Predictably, and unfortunately, the scheme is voluntary; with Russia and India not joining initially. The scheme is also likely to favour the use of biofuel, which is fine if it is made from waste products such as wood chips, but if produced from crops, will have detrimental impacts on land-use and wildlife.

An alternative, or additional, method would be to levy a higher rate of tax on frequent flyers – say anyone making a second leisure flight per year. An even more complicated system would allocate everyone a quota of free air miles per year (a cap which could be raised or lowered) which they could use or sell to others who are willing to pay. This would clearly benefit half the population who are happy not to fly, whilst those who want, or need to fly would be willing to pay more.

The word rationing sounds draconian but is often not as unpopular as critics would suggest, as people see it as being fair to all. If frequent flights became more expensive, then people would more fully appreciate the flights that they do take and would change their habits. A small number of longer holidays would allow more time to relax in the destination and might replace weekend pleasure trips. This would have little effect on cultural exchanges, where people volunteer, work and travel abroad for a few months at a time.

The government could also introduce new regulations to restrict or ban free perks and hidden subsidies that fuel the demand for flights. Many businesspeople benefit tax-free from frequent flyer programmes with points exchanged for free flights for their family's leisure purposes. Some credit cards offer points which can be exchanged for free or low-cost flights. Even the supermarket chain Tesco, which boasts of its environmental credentials, offers air miles linked to expenditure on its credit card. In addition, many quizzes and competitions, often sponsored by business, offer prizes of free flights or international holidays.

****

Three-quarters of emissions from aeroplanes are from short-haul flights. China leads the way in building a high-speed rail network as an alternative to short-haul flights. Other countries should do the same albeit new lines are controversial in densely populated countries like the UK. The French government has banned Air France from competing with the rail network on domestic short-haul routes such as Paris to Bordeaux where the train journey takes less than two and a half hours.

Some companies offer a carrot, rather than a stick, for example they allow their staff to use sleeper trains for business trips, travel first class on long-distance trains, or even offer two additional days annual leave to those who choose alternative means to travel overseas on holiday. Although carrots are attractive in many areas of the climate debate, in aviation they are not enough. Governments will need regulation to persuade people to change their ingrained habits.

The answer from the airline industry to the climate impacts they cause is to continue to encourage growth in passenger numbers, combined with incremental fuel efficiency improvements, experiment with alternative fuels and to offset their remaining emissions.

The problem with alternative fuels is that there are no good alternatives at present. Kerosene is an efficient fuel as it provides a lot of power for its weight and volume as evidenced by the ability of airlines to offer non-stop commercial flights from the UK to Australia. An obvious alternative is biofuel, but to scale biofuel from agricultural crops would require an unacceptably high area of farmland and potentially increase food prices. Household and commercial waste materials can also be used to manufacture biofuel; whilst Finland produces biofuel using pyrolysis, a chemical alteration of wood chips at high temperature. Research is also underway to produce fuel from waste gases, particularly from the steel industry. This sounds promising, but the problem is extra cost, safety, and the difficulty to produce enough volume. Biofuel and synthetic fuels will also not prevent the other climate change impacts of aviation such as producing contrails. However, one interim measure would be for governments to require airlines to use an increasing percentage of alternative fuels, otherwise the industry will not develop them because of their extra cost.

Hydrogen, combined with fuel cells, is another option but given its volatile nature, will require a myriad of technology and safety advancements. In addition, liquid hydrogen is less energy dense and would require more storage space for the equivalent journey than kerosene. Recent trials indicate that electric powered aeroplanes have some potential, but the weight of batteries currently restricts this to small aeroplanes and short flights. Clearly innovation will change these calculations and assumptions in the future. But an important point is that aviation is, for obvious reasons, a heavily regulated and risk adverse industry. It takes decades from early research and development to wide scale deployment in new aircraft, then decades longer to phase out old planes through long industry replacement cycles. Whilst some airlines may invest in the latest more fuel-efficient aeroplanes, there is no net benefit on carbon emissions if they redeploy their old planes to growing markets in low income countries.

Several airlines offer passengers the option to offset their emissions when they book by paying a small premium on the ticket price. Take up by passengers of these voluntary schemes has been abysmally low (3% of Ryanair's customers), but the pressure on airlines to do more is steadily increasing. EasyJet unexpectedly announced in 2019 that it would voluntarily offset all its emissions, mainly through tree planting. Although a small step in the right direction, it may be a rather cynical marketing ploy to encourage its passengers to feel less guilty about flying and to help stave off future regulations on aviation. The offsets do not cover the wider impacts of aviation mentioned earlier, and seem to be too convenient, easy, and cheap – reportedly eight pence per passenger.

Perhaps the only way to fly with a clear conscience is to pay to offset your emissions (doubled to take account of other climate impacts such as contrails) into a scheme that safely and permanently removes carbon from the atmosphere and stores it safely underground. Such schemes barely exist at present but may come into being in coming years. Initially the price will be high; perhaps £200 for a trans-Atlantic flight, but the cost will come down as carbon capture technology improves.

But the concern is that offsetting will distract the public from the adverse environmental impacts of aviation. Even the language of 'offset' makes it sound like you can continue to fly guilt free, when what we really need is to stop carbon emissions. More appropriate language might be for airlines to contribute to a 'climate damages fund', spent on projects to reduce the impact of climate change on people in low income countries. Ideally, the tariff should be set by an international body and set at a sufficiently high level to make an impact. There is more on carbon offsets in a subsequent chapter.

Tackling emissions from aviation is difficult. Until Covid-19, demand was growing faster than improvements in fuel efficiency. Engineers can further improve efficiency using new designs, advanced materials and manufacturing techniques, but there is a physical limit. Governments should set up a massive programme of international collaborative innovation to improve the fuel efficiency of aircraft and explore alternative fuels. They should end subsidies, tax aviation directly correlated to carbon emissions, and introduce a requirement for airlines to source an increasing percentage of alternative fuels. A quota system would be fair on all, but will undoubtedly be politically and practically difficult. All these proposals may need to be phased in as the airline industry recovers from the disruption of Covid-19. They will make flying more expensive and will reduce demand. Aviation and foreign holidays are an area of our lives that we might appreciate more if we place a proper value upon it.

Chapter 20:
Tourism and Leisure

All my five common-sense principles are relevant to tourism and leisure, with 'travel wisely' clearly the most significant. Any discussion of carbon and tourism will be controversial with the tourism industry and the public. Some view any discussion of climate change as a threat to the tourism sector's historical business model of encouraging people to travel further, and more frequently, and to spend more on ever higher quality and luxury experiences.

There is some truth to this, particularly as there is a correlation between the amount spent on a tourism experience and its carbon impact. Think of the carbon impact of a five-star hotel compared with camping; flying long-haul versus a short car journey; or a land-rover safari experience versus walking. Some tourist destinations are inherently unsustainable from an environmental standpoint. Las Vegas and Dubai have built a reputation as international tourist destination. Both are in desert regions yet use excessive amounts of energy and water for hotels, swimming pools, irrigated golf courses and have outdoor fountains as visitor attractions. Las Vegas gets its water from the drought-stricken Colorado River, whilst Dubai sources its water from energy intensive desalination of sea water. Dubai, in the hot desert, also boasts the largest indoor ski centre in the world.

Globally, tourism is growing and already accounts for 8% of human carbon emissions so we need to face up to it. In most high income countries, the number of bank holidays and paid holidays has steadily increased for decades although citizens of the USA get fewer paid holidays. With increasing affluence people take more breaks, longer holidays and choose to travel further. Meanwhile, the number of tourists within and from China has exploded in the last decade.

The fundamental issue is flying, discussed in the previous chapter, which needs to be controlled until airlines can introduce new low emission technology or until a secure and permanent carbon offset scheme is established. The implication for countries like the UK is that there may be a reduced number of overseas visitors, but more British people would choose a staycation and spend money within the UK. British people take 72 million trips overseas each year spending £48 billion; whilst Britain welcomes 38 million overseas visitors spending £28 billion - so any restriction on flying would bring a net benefit to British tourism.

Many locations, regions and even entire countries have built their economy around attracting visitors who fly in for one or two weeks. Many remote islands have few other opportunities for economic development. Even if the attractions and accommodation provided are eco-friendly, visitors' flight emissions will still have a significant impact. There is no easy solution, but having countries and islands reliant on long-haul tourism is not a sustainable answer.

Tourism can be a passive and relaxing activity, or it can be an active experience that broadens the visitors' cultural and educational understanding of the world. An argument can be made to support tourism if it brings wider socio-economic benefits, particularly, if it involves local communities. In Africa overseas tourists on wildlife tours bring in income to help manage national parks. Without these visitors there would be fewer jobs for local people to protect nature and wildlife. In the wildlife haven of the African Okavango Delta, the local community provides the guides to take tourists in canoes through the marshes to community owned campsites or lodges. This is a good model to support the local economy and it provides locals with a vested interest to protect the landscape and combat agricultural encroachment and poaching. Meanwhile, local experts can teach visitors about wildlife and culture.

Some young people volunteer with overseas charities. If well organised, such opportunities can have multiple benefits. The young person will learn about other cultures, develop new skills and friendships, will do some good in the country during the trip, and they are then more likely to engage in a lifetime of supporting international organisations to protect wildlife.

The possible 'solutions' to international aviation are more thoroughly considered in the previous chapter. For tourism involving long-haul flights, the common-sense compromise is that people should treat such trips as an occasional treat, for long lengths of holidays. Long-haul trips should not become a routine or regular choice.

****

Ships use a lot of energy, usually heavy fuel oil, to propel themselves forward as water is a dense fluid compared to air. Cruise ships due to their size are the biggest emitters; but ferries, marine wildlife cruises, motorboats and activities like water skiing are carbon intensive too. Cruise ships are like mini cities on the move and have a surprisingly high level of carbon emissions per passenger. Cruise ships have been increasing in size leading to citizens of cities like Venice campaigning to ban them because of too many visitors in a short space of time that do not spend money on accommodation. As a starting point, government should require all cruise companies to publish the carbon emissions per passenger of their ships to help people to make better choices. Furthermore, in addition to government taxes on the fuel burnt, a levy on passengers should be collected and spent on collaborative international innovation and investment to research and develop the more efficient technology needed to reduce the carbon emissions of cruise ships.

Over time new technology such as electric hybrid ferries, computer operated sails and better streamlining will gradually reduce the emissions from ships, but at present they are not a low carbon form of transport. Again, we must make difficult choices and balances.

****

Hotels and other accommodation providers are big energy users, buy large quantities of food and most discard high amounts of food waste and packaging. The tourism sector can tackle energy if the will is there by investing in energy efficiency and on-site renewable electricity generation – many tourist hotels are in sunny locations or near a windy coast. Hotel owners and restaurants have a big influence on the food choices that they offer to their customers. Buying locally grown fruit and vegetables in season will help the local community to benefit from the influx of tourists, but the most significant impact will be to offer more vegetarian and vegan options to their guests. Hotels can tackle food waste by offering smaller plates, or more choice of serving size. In theory, buffets, where guests select their own food can reduce food waste, but they usually result in leftover food at the end of the serving period and they may encourage guests to eat and drink to excess. Care can be taken to avoid excessive food packaging, for example individually wrapped butter, jam, and yoghurt tubs. Other minor issues can easily be tackled. Do guests need so many pillows and towels? Do they need a new soap bar every day wrapped in plastic, or new shampoo bottles for every guest?

But despite some great initiatives by some hotels and hotel chains, most people select accommodation based on its location, the amenities on offer, price and brand reputation rather than on how sustainable the accommodation is. For this reason, yet again, we need more regulation. A small example is the Scottish Government's regulations that require commercial premises including hotels to collect their food waste separate from other waste. It can then be composted or taken to an anaerobic digester for energy recovery.

Given that most holidays involve travel and an increase in normal levels of consumption, an obvious, but not universally popular, solution to the carbon emissions from tourism is to take fewer, but potentially longer breaks. Enjoy better-quality holiday experiences, avoid frequent short weekends away and reduce or eliminate your flights. This is tied up with how satisfied we are with our daily lives - our work, homes, friends, local leisure pursuits and our local community. Can we enjoy spending more of our free time at home, or do we feel the constant need to travel and 'get away'? Improving our local neighbourhoods with more green space, cycle lanes and opportunities to walk would help.

Obvious choices for low carbon holidays include walking, cycling, golf, sailing, sunbathing, sightseeing or camping. Holidays by (electric) car with a family of four sharing one car can be efficient. Similarly, a tour with fifty people in one coach will result in low emissions per passenger. However, care must be taken choosing 'green' holidays, as some places marketed as 'eco-lodges' may still result in excessive emissions. Although they may be visually attractive and energy efficient, they are often wastefully large, offer energy intensive extras such as hot tubs, or are in remote locations with extensive travel to reach the accommodation and to visit attractions during your stay. Large hotels, which have invested in energy efficiency, may be more eco-friendly than separate accommodation for individual families.

Some tourist attractions successfully double up as environmental educational facilities. In England, the prime example is the Eden Project in Cornwall, a redeveloped abandoned china clay pit. Over one million people visit each year to see the two huge glasshouses offering visitors the experience of all the world's main natural biomes in one location. Its main aim is to educate children and adults, but in a fun and interesting environment. They have an active procurement policy to buy local, with over 80% of the food they serve sourced from Cornwall or Devon, much of it grown within their own nursery.

In Scotland, examples include the Whitelees Wind farm visitor centre, Pitlochry hydro-electric dam and fish ladder, the Scottish Seabird Centre at North Berwick, Glasgow Science Centre and Dynamic Earth in Edinburgh. Nature based tourism such as wildlife tours also have a useful educational element to them. Iceland now makes more money from whale watching tours than from hunting whales for meat.

****

Our society is not likely to ban all high carbon holidays and leisure activities - although I would make the exception of space tourism and ban it before it takes off! So, how do we nudge people towards making better choices and encourage the industry to offer better choices?

The Green Tourism scheme is a UK industry led initiative to encourage tourism businesses to be sustainable. An independent assessor assesses their sustainability credentials and awards gold, silver or bronze status. Such schemes do raise standards but will only be truly effective if consumers use them to influence their decisions on choosing accommodation and what attractions to visit. The evidence suggests that these voluntary schemes have a limited impact.

The government should target taxes, subsidies and incentives to encourage all building owners, including hotels and leisure attractions, to invest in on-site renewable energy and energy efficiency. Some cities and local authorities have introduced local tourist or bed taxes with the tax directed to improve the local environment and services to benefit locals and tourists alike - to improve public transport, build footpaths, fund free art or open-air concerts, or simply to clear rubbish left by tourists. Other regions, such as Mallorca, are encouraging more sustainable tourism with a focus on smaller numbers of high-end visitors rather than large numbers on a tight budget which can overwhelm a local community.

Governments could introduce tax breaks for community run tourism accommodation and attractions, and to privately owned tourism facilities that enter a strong partnership agreement with the local community. Partnerships may offer up unexpected and novel ideas. They could be between accommodation providers, tour companies, travel providers, the local community, a local farm or visitor attractions. An example is a rail operator collaborating with a tourist attraction to offer combined discount tickets for travel and entry. Many cities offer discounts on public transport for visitors. Smart technology on mobile phones and combined travel cards can enable this collaboration.

More widely, government funded industry bodies should shift their marketing budgets away from weekend breaks and attracting small numbers of wealthy tourists from overseas, towards initiatives aimed at the European and domestic sector. They should promote holidays and attractions that are lower carbon options, for example, travel by train or holidays that focus on relaxation or sightseeing.

Meanwhile, businesses can provide visitors with information on how to use public transport, provide free or subsidised bike hire or information on local walking routes. Councils can ensure there are enough public electric charge points to enable holidays by electric car, and hotels and visitor attractions can provide their own charge points.

So, tourism is well-placed to reduce its environmental impacts but the problem of long-distance travel, particularly long-haul flights remains. In tourism, we can make a difference with the individual choices we make.

****

Mirroring the growth in tourism, many of us have more spare time and money to pursue our regular and occasional leisure activities and hobbies. High carbon leisure activities are likely to incorporate one or more of the following: travel, motorised boats, the need to buy new equipment or clothing, or a constant pressure to upgrade to the latest fashion or equipment.

Low impact leisure activities are much the opposite; those that require little travel, avoid motorised equipment and have no requirement to regularly buy new clothes or sports equipment. Exercising near home has a low impact. Other examples include dancing, walking, card games, cross-country skiing, art classes, tennis, judo, bingo, yoga, reading, and consensual sexual intercourse if this does not lead to pregnancy!

Many activities like cycling, hill-walking, orienteering, golf and fishing can be low or high carbon, depending on how far people travel to participate in the activity and how frequently they buy new equipment. It is a simple calculation, but if you share a car with three other people to take part in an activity then the carbon impact per person will be one-quarter than if you all drive separately.

The carbon impact will depend on how and where you play or watch your chosen sport or activity. Football played in a local park can be low impact. In Malawi, children play in bare feet using a ball cleverly made from used plastic bags. Compare this with playing for a club football team, regularly buying new equipment, driving to regular weekend away matches and flying on the occasional overseas tour. Similarly, for spectators, the carbon impact will depend on how and where you watch your chosen team. Watching a home football match is low impact compared with a fan following an English Premier team not local to where they live. They need to travel even to attend 'home' matches and will increase their carbon footprint by flying to watch regular European matches.

For a big global event, travel by participants and spectators is the most significant component of the event's carbon footprint — 86% in the case of the football world cup held in South Africa. Meanwhile, to celebrate 60 years of the European Championship, it was planned to hold football games in the postponed Euro-2020 championship, in twelve different countries. Supporters of Poland were expected to travel 6,000km in ten days to Dublin, Bilbao in Spain and back to Dublin just to see the group stage. Was the travel and carbon implication of the decision to hold games in 12 countries properly taken into consideration?

Downhill skiing is an activity that requires a lot of energy use. Many Britons fly to the Alps for a regular ski holiday, creating emissions from the flight, transfer to the hotel, accommodation, nightlife and mechanised tows to take them up the mountains. Much of this energy is sourced from renewable sources (or nuclear in France) but this means that this energy is not available for other purposes. Due to climate change, many low-lying resorts have already closed, and most resorts are becoming increasingly reliant on energy intensive equipment to make snow. The irony is that skiers use energy which contributes to climate change and are therefore destroying the basis for their hobby.

Leisure activities also encompass the arts and culture – cinema, theatre, music, concerts and museums. Apart from the energy used to heat the buildings that house these activities, cultural events tend not to be as carbon intensive as more mechanised forms of entertainment. The exception is if you travel long distances to see the culture, for example, travelling to Italy to view famous buildings and works of art. Like football matches, the travel to a concert is the largest component of its carbon footprint. Having good quality and accessible public transport, whilst actively discouraging people to travel by car, is the best policy to reduce the carbon footprint of such events. Organisers often put on coaches to get young people to music concerts in remote rural locations. The Solheim Cup golf event at Gleneagles in Scotland is a good example of a major event held in a rural area where no public car parking was provided. Instead, visitors had to travel by train or use the park and ride facilities set up for the event.

The creativity of people who work in art and design needs to be harnessed to challenge, inform and engage audiences on the impacts of climate change and in the actions we can all choose to take to reduce our carbon footprint. This can be active, for example, an exhibition on climate change or embedded more subtly into wider messages. For example, the BBC's Blue Planet has had a huge impact on raising awareness across the world of the impact of plastic in the oceans. The film Avatar appealed to a wide audience of science-fiction fans but contained a strong pro-environmental message. Even computer gaming can be transformed from morally dubious shooter games to creative and educational games. Minecraft is a game where you build new cities and virtual worlds with building blocks. They have launched a fun, but educational, module for players to build cities powered by sustainable energy.

****

Keeping a pet can have a significant carbon impact, particularly meat-eating carnivores like cats and omnivores like dogs. Herbivores, like birds and rodents, have less of an environmental impact. Americans keep 78 million dogs and 86 million cats, that between them consume as much food as 60 million Americans, emit 60 million tonnes of carbon and produce over five million tonnes of faeces each year. Whilst American people eat a diet containing around 20% meat, their pets' diet is over 30%. Some of this is 'waste' meat that humans are not likely to choose to eat, but a recent trend is for people to feed their pets higher quality meat that could be fed to humans, and of course, many pets are overfed.

Meanwhile, cats, as the most common carnivore, have a devastating impact on wildlife. In Australia, 3.7 million domestic cats kill 230 million native reptiles, birds and mammals a year, plus a further 150 million introduced animals such as rodents. In addition, an estimated 2.1 million cats have gone feral, killing a large but unquantifiable number of animals and birds. Cats can devastate ground nesting birds - particularly on islands where birds evolved in the absence of cats. In New Zealand, cats are responsible for the extinction of six endemic bird species and 60 sub-species. In Scotland, feral cats have bred with the native wild cat diluting its pure genetic diversity and leading to a weaker hybridised species. We should do more to ensure all pets are cared for, neutered if appropriate, and owners should keep cats indoors or wear a bell to warn wildlife of their approach.

The number of pets kept around the world is increasing, as people in fast developing countries like China decide to keep pets for the first time. However, keeping pets can have its uses. Some dogs are working sheep or guard dogs, some cats provide a pest control service, whilst all pets can be good for companionship and mental health. Walking a dog provides good exercise for humans.

****

We occupy an increasing proportion of our spare time with using internet enabled electronic devices whether to stream music, films, TV programmes, computer games or social media. As our digital connectivity and use of the internet grows so does its impact. The total emissions from business and leisure use of electronic devices are already larger than from aviation, using around 10% of global electricity and causing 4% of carbon emissions. Despite energy efficiency improvements, the total impact continues to increase. The emissions are from electricity; split roughly one-third between charging or powering our devices; data centres that store information; and the network to transfer data to our homes, businesses and devices.

Using Wi-Fi is more energy efficient than using 4G to access the internet, whilst 5G will allow us to all access far more data therefore further increasing energy consumption. Similarly, using high definition television and video increases data usage, and the trend towards connecting more objects to the internet will all increase power consumption. Consumers can reduce this by avoiding unnecessary use of high definition, for example, there is no need for it on small mobile phone screens, and there is often no need to be streaming video at the same time as listening to music. Better alternatives are to listen to music only on sites such as Spotify or switch to low definition video.

Our holidays and leisure activities are an area of our lives where we have a lot of freedom to make choices. How many of us enjoy airport security, customs, passport control and cramped seats? It would be better to take most of our holidays locally, saving the airport experience for occasional longer, special trips. How we participate in our leisure activities is often more important than our choice of leisure activities. Attending local events is the key to reduce our travel impact, followed by using public transport or maximising the opportunity and sociability of car sharing. Reversing the peer pressure to continuously upgrade to the latest fashion or newest equipment would lower our carbon footprint and reduce the cost of participation which would enable a wider audience to access sports and activities.

Chapter 21:
Agriculture

Of my five common-sense principles, the most relevant to agriculture is to 'nurture nature', followed by 'avoid waste' and 'price carbon pollution'. Meat and dairy form a key focus of this chapter because of their disproportionate environmental impact, but other animals and crops are considered too. The next chapter considers our dietary choices and later chapters on nature and carbon offsets consider the wider issues of rural land-use.

Land-use change and agriculture directly account for one quarter of global greenhouse gas emissions, over half of which are from agriculture. Emissions from agriculture include nitrous oxide from managing manure and using synthetic nitrogen fertiliser; fossil fuels from farm machinery and transport; and methane from cattle, sheep and rice paddy fields. The remaining emissions are accounted for by permanent alteration of land-use from forest or grassland to agriculture with a consequent loss of carbon stored in trees and soils - carbon accounting in this area is complex. But the greenhouse gas emissions from our diets are even larger than from this use of land. There are also emissions from manufacturing fertiliser, transport, food processing, packaging and from cooking our food. There are a range of estimates on the total emissions from what we eat, but one quarter is in the right ballpark.

Agriculture takes up half of the world's habitable land area and accounts for 70% of all freshwater used, mainly to irrigate fields. The expansion of agriculture and pasture continues to be the greatest global pressure and threat to wildlife and biodiversity. Meat and dairy products create half of these agricultural emissions - as much as from all cars, planes, trains and ships combined. The global carbon impact of cattle is greater than from every individual country in the world except for China. Due to ongoing consolidation in the sector, the top 13 global dairy companies emit as much carbon as the UK, the 6th largest economy. Emissions from dairy continue to rise due to increased demand outstripping efficiency gains.

Globally, we grow more than double the amount of food that humans need to eat healthily; yet 800 million people are undernourished because of politics, social policies and distribution difficulties. Despite this, the average person eats more calories than they need - our metabolism burns most of this excess off as waste, whilst a small proportion affects our waists, causing obesity. Of the total crops produced, at least 20% is wasted and never eaten, 15% is converted to biofuel and 30% fed to livestock – leaving 35% directly consumed by humans. This 30% of food fed to livestock, combined with calories direct from grass and pasture, result in meat and dairy only providing 20% of the total calories that we eat. As an example, we can eat soya beans, convert them to biofuel or feed them to animals. In the latter case, only one-tenth of the energy is converted into calories for human consumption. The warm-blooded animal uses the rest to keep itself warm or wastes it as manure or methane. Another statistic that demonstrates the inherent inefficiency of livestock farming is that it occupies 77% of total farmland but only produces 18% of the calories that we eat. Admittedly, much of this pasture is less naturally productive land, but we could use some of it to grow food direct for human consumption, for commercial forestry or it could be kept aside for wildlife.

Cattle, sheep and goats are ruminants. A ruminant is a mammal that ferments food in its stomach prior to digestion and in the process emits methane. Every cow burps 250 to 500 litres of methane a day. Scientists commonly quote methane as being 28 times as potent a greenhouse gas than carbon dioxide, but this is over a standard 100-year period. Unlike carbon dioxide which can remain in the atmosphere for centuries, methane breaks down in the atmosphere after a decade. It has an impact 84 times that of carbon dioxide in the next critical two decades in the fight against climate change. Recent trials show that it is possible to reduce, but not to eliminate, these emissions through selective breeding programmes, or at extra cost, by feeding livestock a higher fat diet or adding selected feed additives to their food. Given the scale of methane emissions from ruminants, further research and innovation would be beneficial. However, it seems unlikely that science and technology will make a transformational impact. For example, Cargill is a large American corporation that provides agricultural and industrial products. They process eight million cattle per year, mostly fed on grain. Under their BeefUp sustainability initiative, they collaborate with farmers to use less fertiliser, reduce ploughing, improve manure management and restore soil fertility. They also research new feeds and selective breeding. Dairy cows must produce a calf each year to continue lactation - Cargill are crossbreeding beef and dairy cattle to produce dairy cows whose calves can be removed and fattened more efficiently for beef. Despite all these initiatives, their target is only to reduce greenhouse gas emissions across their cattle supply chain by 30% by 2030.

****

There are numerous other environmental and ethical impacts from raising the 1.5 billion cattle on our planet, particularly acute from industrial farming of large herds in confined spaces. Cattle grazing outdoors require large areas of land, whilst some animals reared intensively never graze on a blade of fresh grass. Much of the destruction of the vast Amazonian rainforest has been to grow soya to feed cattle or to clear land direct for pasture. In nature, manure is a natural product that fertilises the soil, but it is concentrated in intensive livestock rearing. If not carefully managed, it has the potential to pollute groundwater and rivers as its nutrients feed algae which quickly multiply, use up the oxygen and kill fish. Processing manure releases ammonia, methane and nitrous oxide. Ammonia can react with sunlight to create smog and contribute to acid rain. However, good management of manure, using anaerobic digestion for example, can reduce odour, create biogas and a fertiliser from the residue. Subsidies and advice to promote good manure management and energy recovery will help.

As always there are complications, and the devil is in the detail. In the worst case, farmers raise cattle on land converted from tropical forest, use concentrated feed and transport the meat long distance to market by ship and lorry. In semi-arid lands, growing fodder and drinking by cattle can strain local and regional aquifer water supplies, and put cattle ranchers in direct conflict with wildlife over limited water resources. Beef raised intensively in North America on huge feed lots, primarily eating local grain and given growth hormones is perhaps unethical, but relatively efficient from a carbon perspective. Likewise, beef raised in the damp conditions of the UK could have half the carbon emissions of beef imported from tropical countries like Brazil. Our productivity is high, with fast growing rain-fed grass, and the soil can store carbon, at least up to a point. But grass-fed cattle grow more slowly, and therefore live longer and emit more methane before slaughter. So, UK cattle farms have high welfare standards, with few reported pollution incidents and emissions are lower than in most other countries. But despite all of this, beef and dairy farming, regardless of source, is still a high carbon emitting activity compared to most plant-based foods. We need to halt the continuing increase, and then reduce the global number of cattle and sheep. In future, a limited number of cattle and sheep would be best kept in certain landscapes to support specialist plants and wildlife that are adapted to their grazing.

****

The environmental and health issues arising from industrial rearing of livestock and poultry extend beyond land-use and carbon emissions. Three-quarters of all the world's antibiotics are fed to livestock and poultry often to reduce the risk of disease spreading amongst animals and birds kept in large numbers in confined spaces. Residues of chemicals such as worming treatments remain in manure, can leak into the groundwater and some remains in the meat that we buy although cooking should make this safe to eat. Cattle dung can poison insects such as dung beetles which would naturally process it into soil nutrients.

Despite all these precautions, there are numerous diseases and impacts on our health that can arise from our intensive rearing of livestock and birds - salmonella, E-coli, listeria, foot and mouth disease and BSE. Then there is the scary, but real issue, of the spread of antibiotic resistant bacteria, caused by the excessive use of antibiotics in animals and humans. The risk is that bacteria become resistant to the antibiotics that doctors prescribe to patients which would result in routine operations becoming very risky and deaths arising from minor infections. This could quickly become a global health issue.

Every time a new outbreak of disease occurs, governments feel obliged to pay compensation (a subsidy) to support farmers and the meat processing industry during the outbreak to ensure that everyone in the food chain complies with strict rules and to prevent further spread of the disease.

The UK identified BSE, popularly known as mad cow disease, in cattle in 1986. Initially it was not treated too seriously because of the 20-year incubation period between eating infected beef and symptoms of brain damage appearing in humans. BSE seems to have arisen due to cattle, a herbivore, being fed an unnatural diet of bone meal. To halt its spread, farmers had to incinerate over one million cattle during the peak years of 1992-96. The BSE enquiry concluded that the disease had cost the UK taxpayer around £4 billion to prop up markets, implement slaughter schemes and to compensate everyone in the meat production chain.

Bovine TB is an endemic infectious disease that spreads between cattle, and between cattle and wild animals such as badgers. The UK Government funds the TB Advisory Service to advise farmers on biosecurity measures. In 2016, farmers slaughtered badgers and more than 29,000 cattle to try and control the spread of the disease, with taxpayers spending £100 million per year to support farmers.

An epidemic of foot and mouth disease in sheep and cattle hit the UK in 2001. Selective slaughter and strict quarantine are the only ways to combat this highly infectious viral disease. Farmers slaughtered six million cattle and sheep at a cost to the taxpayer of £3 billion with a further £5 billion cost to the private sector. In this case, the impacts spread far beyond the agriculture sector to tourism and hospitality businesses as the government asked people not to visit the countryside to avoid the risk of spreading the disease.

The UK has now learnt some lessons from these catastrophic disease outbreaks by implementing much stricter controls, such as the ability to trace all animals from farm to slaughterhouse.

Meanwhile, the EU paid Namibia to erect fences to try and stop the spread of foot and mouth disease between cattle and wildlife. Although well intentioned, these fences were erected quickly without any research on animal migration routes. Tens of thousands of wildebeest slowly starved to death as access to their seasonal migration routes was abruptly blocked.

****

Like cattle, sheep are also ruminants. Unlike cattle, most sheep graze on pasture that is not in competition with other agricultural uses, although we could convert some of it to forest or manage it to benefit wildlife. Most of Britain's uplands remain treeless because of the impact of grazing animals such as sheep, deer and rabbits that prevent new saplings from growing. Sheep farms dominate the picturesque Lake District and uplands of Wales and tourists photograph these landscapes as a green and pleasant land. But none of this is natural. In the past, farmers in the EU were paid a subsidy for each sheep kept resulting in overgrazing. Wildlife would benefit from reduced grazing pressure, and for subsidies to shift to farmers becoming active environmental stewards of our countryside.

All intensively reared livestock, caged birds and fish farming cause significant environmental impacts. The carbon emissions from industrial farmed chicken and pork are around one third of those from beef and mutton but are still higher than that of most plant-based foods. The number of chickens in the world has doubled since 1990 to 24 billion, mostly kept in intensive chicken or egg farms. Most are kept indoors in crowded conditions, leading to the risk of diseases such as salmonella and the consequent need to use antibiotics. Their manure leads to emissions of ammonia and the risk of water pollution.

Fish farming is another intensive method to grow food that uses antibiotics to reduce the risk of disease, and the excess nutrients in the fish waste pollute the surrounding waters. Although, carefully regulated fishing from the sea can be sustainable, globally, the current industrial scale of fishing is unsustainable. China has by far the largest fishing fleet, with huge industrial scale boats accessing distant waters that are often unregulated or where regulations are poorly enforced. Industrial fishing usually results in non-target species such as sharks, being caught, and thrown back dead into the sea, either because there is no commercial value or because it is a regulatory condition not to catch that species. Most governments subsidise diesel for fishing boats - the economics of fishing across the globe are only sustained by these subsidies – and one-fifth of the global catch of wild fish is fed to livestock or is ground up to feed fish in fish farms.

It is the industrial intensification and overcrowding of livestock and aquaculture that leads to many of these environmental impacts. Eating animals, birds and fish that live free range or in the wild has a much lower environmental impact, although perhaps with the exception of wild fish, the supply of such meat is limited and more expensive, and care must be taken not to overexploit the resource.

****

Just in case vegan readers are feeling complacent, the climate and environmental impact of food is much more complicated than simply meat versus plant based. Rice can be a high impact food, estimated at 6% of all agricultural emissions. Bacteria produce methane in water-logged soils such as flooded paddy fields. Many countries grow rice with excessive, subsidised fertiliser which emits nitrous oxide, a gas 300 times worse than carbon dioxide for the climate. Farmers can reduce methane emissions by shortening the period that they flood fields and by adding sulphate to fertiliser which outcompete the methane producing bacteria. Given the vast scale of rice grown, in different soil types and climates, we need more research into optimal crop and water management techniques followed by new regulations and advice to farmers.

Many water-stressed countries use their limited water supplies to grow crops for export. Avocados contain potassium and fibre and have become popular as a health food, but each requires 70 litres of water. Avocado farmers in California have had to abandon some agricultural land because of drought, whilst in Chile wealthy farmers have bought the rights to access water to irrigate avocado trees. They have diverted water from rivers, and during drought the water-table has fallen drying up some rivers and forcing villages to tanker in all their water supplies. Drip feed agriculture, using technology perfected in Israel, can reduce water consumption, but whatever is done, many countries will continue to choose to grow crops for export income at the expense of some local people. It would be more equitable for the domestic supply of water to take priority over agriculture for export.

Almonds are California's most valuable export crop, benefitting from the climate and new mechanisation techniques to harvest them. Almond trees consume around 10% of all California's water, some pumped from groundwater, depleting the aquifers. The trees depend on bees to cross-pollinate to create the seeds. Intensive agriculture has decimated habitat for native bees so farmers import domesticated honeybees. However, this trade can transfer infections, and large honeybee colonies can outcompete native bees. The colonies are also prone to sudden collapse, perhaps due to pesticides and infections from pathogens.

Tractors, harvesters and other farm machinery run on diesel. Their heavy weight makes them harder to electrify, but manufacturers have made little effort because there has been no government support, and in many countries diesel for farm vehicles is subsidised. The solutions to this are obvious; clear government policy with targets placed on manufacturers backed by a phase out of subsidies. Given that most farms have the space to generate their own renewable electricity, the benefit to farmers of electric or fuel cell vehicles are clear.

There are many who campaign for organic produce and argue equally passionately against genetically modified foods. Organic farms only use organic fertiliser, natural pesticides and use techniques like crop rotation, mixed crops and encourage natural insect predators. They ban synthetic fertiliser, pesticides; and growth hormones and routine antibiotics in livestock. Organic farming methods are usually beneficial to wildlife, particularly the reduction in herbicide and pesticide use, and it often goes hand in hand with a farm that manages its soil better and has more land set aside that is not intensively cultivated. Insects, such as pollinating bees and moths, can thrive in such a landscape. Reduced fertiliser run-off into rivers improves water quality.

But organic farming usually produces a lower yield than intensive agriculture using synthetic fertiliser. So, a counterargument is that moving to organic farming would require more land to produce the same amount of food with negative consequences on the area of remaining land available for wildlife. Despite, using no synthetic fertiliser, greenhouse gas emissions can be higher because of this lower crop and livestock productivity.

The debate is polarised. But there may be a sensible compromise between the two extremes of organic and intensive agriculture that will avoid the overuse of synthetic fertiliser, maintain long-term fertility of the soil and feed a growing population. Regenerative agriculture is one such compromise. This is an approach that aims to enhance the soil, land, water quality and wildlife. The techniques are to minimise tillage, always keep soil covered by plants, plant a diversity of crops and possibly to integrate livestock into arable farming. These will increase the micro-biology in the soil making it naturally more fertile, reduce soil compaction, and consequently improve water quality downstream through less erosion and nutrient run-off. Although it has similar aims to organic farming, the criteria are not so strict. Regenerative agriculture focuses more on reducing expensive inputs to fields and avoiding ploughing. This reduction in cost will boost farm profits. General Mills, the American food processor of ice-cream and the breakfast cereal Cheerios, is investing in pilot schemes for regenerative agriculture. This is a partnership with farmers, the municipal water company and a university to monitor water quality and carbon sequestration in the soil. They employ consultants to engage with farmers and to check the effectiveness of the trials, but in the long run such ideas spread fastest through government information and subsidy schemes or by farmer to farmer peer contact.

The James Hutton Institute is a Scottish led research agency that promotes sustainable management of land and crops. They own a lowland and upland farm where they pilot good crop and livestock practice and measure its impacts on soil, water quality and carbon emissions. Across the world we need more research into the whole complicated area of soil health and best farming practice then disseminate this to farmers.

Genetically modified (GM) technology is the deliberate modification of an organism's DNA by transferring genes from one organism into another. For centuries, farmers and scientists have used conventional techniques to breed more productive crops and animals, but GM is a further, more complex and intrusive step that can lead to companies patenting new crops. To date, most GM crops have been bred with herbicide tolerance and insect resistance traits which reduce the need to spray herbicide and pesticide. This low tillage farming increases carbon content in the soil and reduces the need for heavy farm machinery - avoiding soil compaction. An added benefit is that the food that we eat will contain lower residues of insecticide and herbicide. New GM crops are now being developed with more proteins, carbohydrates or vitamins with the aim of enhancing nutrition and human health. Plants can also be designed to be hardier, and to grow better in dry, hot or cold regions.

Many environmentalists dislike the whole concept of what some call Frankenstein crops. The EU has banned growing GM crops. The arguments include the risk of cross-pollination with other native plants, the development of pesticide resistant pests due to GM's reliance on a limited range of pesticides, and the risk that genes could transfer to non-GM crops creating herbicide resistant weeds. Many also dislike the idea of our agriculture becoming dependent on multi-national life-science companies, with new seeds dependent on a small number of seed suppliers. Some are legitimate concerns, although there is no scientific evidence that eating GM foods is harmful to human health.

On balance, the environmental arguments in favour of GM crops are strong, with increased output and less need for herbicide, pesticide and insecticide. However, we must carefully regulate the GM industry for environmental reasons and to reassure the public. The GM industry does not require any public subsidies as successful multi-national companies already lead the industry.

****

In many countries, particularly China and India, subsidies are available to farmers to reduce the price of synthetic fertiliser leading to their over-use. These should be phased out except in countries that suffer from malnutrition. Going further, governments should tax synthetic fertiliser to take account of its emissions of nitrous oxide, and tax cattle and sheep to take account of their methane emissions. In the latter case, the tax could apply to the farmer, or the consumer buying meat and dairy products. These proposals are unlikely to win a popularity competition with farmers, but clever redirection of subsidies would mitigate any overall affect. A tax on fertiliser would reduce its use, but if implemented unilaterally by one country would make that country's products less competitive on the global market. Compensation, even as simple as a direct payment to farmers could make up for this, paid out of the tax income raised. Farmers could spend this on more fertiliser but are equally likely to spend it on other priorities with lower carbon emissions. In addition, a small tax on buying these products could, like the UK's sugar tax on fizzy drinks, have a disproportionate impact on behaviour by sending a clear message to consumers about the carbon impact of these foods.

We should abolish subsidies which encourage wasteful use of resources. In the UK, one example would be the red diesel concession where farm machinery, tractors and refrigerated trucks pay a lower rate of diesel fuel duty. These subsidies, whilst well intentioned, discourage the necessary innovation and investment in new technology which is necessary to reduce local air pollution and carbon emissions.

Governments should completely redirect agricultural subsidies away from encouraging maximum production to a payment to farmers to undertake actions that will benefit the environment and restore ecosystems. A prerequisite, before being eligible to receive any subsidy should be for farmers and landowners to undertake a greenhouse gas audit of their operations, taking account of different soil types, fertiliser use, crops and livestock. This basic information is necessary to support good land-management decision making. Although this proposal to restructure subsidies towards helping the environment applies world-wide, the UK Government has already published plans to implement similar measures after leaving the EU. Such an action can lead to multiple benefits; for example, planting trees along riverbanks can protect soil and nutrients from being washed into rivers and the shade can lower the water temperature to protect fish from warming temperatures. Reduced fertiliser use, and careful application when needed, will be more resource efficient, reduce air pollution and improve water quality. Planting hedges is good for wildlife and provides shelter to livestock.

Farming has gone through several innovation-led revolutions over the centuries. In Europe in the 18th century, the Agricultural Revolution increased crop productivity by introducing new machinery, crop rotation and planting crops like clover or turnips which fix nitrogen into the soil. The mechanisation of farming occurred in high income countries in the first half of the 20th century. Tractors and combine harvesters replaced horses and manual human labour. The Green Revolution in countries like India, in the second half of the 20th century introduced high yield varieties of wheat and rice, along with increased use of synthetic fertiliser, pesticide and irrigation.

Technology continues to increase agricultural productivity by increasing output and reducing labour. Drip-feed irrigation, improved crop varieties and GM crops are recent innovations; with new initiatives such as indoor, vertical agriculture now being trialled. There is nothing to suggest that this continuous process of innovation and change will stop. The next innovations, which have the potential to radically disrupt our agriculture, are the fermentation of proteins to grow meat substitutes and growing artificial meat in laboratories. Given the pace of technological advance, the urgent need to reduce the environmental impact of meat production and the latent demand from many consumers to eat less meat; it is likely that the meat and dairy agricultural landscapes that we currently know will change radically over the next decade or two. A later chapter will explore this opportunity to rewild our lands to improve our lives, the environment and to benefit nature. In the meantime, a shift to regenerative agriculture will reduce carbon emissions; whilst less ammonia and nitrous oxides will improve air quality, soils will contain more living and organic matter, and groundwater, rivers and estuaries will be cleaner.

Chapter 22:
Our Dietary Choices

This is a challenging subject as food and eating habits directly affect us all, and one that many feel strongly about. Our food choices have a significant impact on carbon emissions and on wildlife through changes in land-use. Yet the price we pay for food does not take account of the methane emissions from cattle or nitrous oxide from fertiliser, so our choices are not aligned to the environmental damage caused. "Listen to the science" said Greta Thunberg. Well the science clearly points us towards using less artificial fertiliser and to eat not just vegetarian, but to adopt a predominately plant-based diet. This advice may be controversial with some because they enjoy eating meat, dislike or distrust anyone advising them on what they should or should not eat or they have an emotional attachment to existing farming and landscapes. All I can suggest is to try and put aside any prejudice and vested interests and read on.

Of my five common-sense principles, the most relevant is to 'nurture nature', followed by 'avoid waste', 'price carbon pollution' and 'consume carefully', meaning for many of us to eat less.

The chart below shows the average greenhouse gas emissions of some food types. It includes the emissions from the farm (66% on average), global land-use changes (20%) and a smaller component from transport and retail (14%).

Figure 2: Greenhouse Gas Emissions of Food – global average

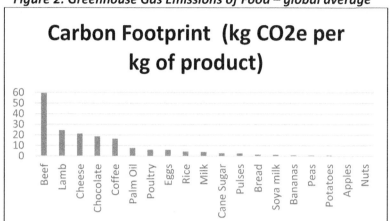

Carbon Footprint (kg CO2e per kg of product)

Source: Poore and Nemecek (2018), OurWorldinData.org

Foods that cause high carbon emissions to produce include beef, lamb, cheese, chocolate, coffee, and palm oil. Foods with relatively low carbon emissions include plant-based proteins such as nuts, peas, beans, lentils and tofu; fruit such as apples and pears; vegetables such as carrots and potatoes; and staple crops like maize and wheat. Beans, pulses, peas and lentils absorb nitrogen from the air into the soil and therefore require less artificial fertiliser than other plants. The differences between foods can be startling. Producing one kilogram of beef emits around 60kg of carbon, whilst growing the same weight of pulses emits only 2kg. Overall, the carbon impact of a vegan diet is around half that of a regular meat eater. Choosing to change your diet could have a similar impact on your overall carbon emissions than giving up your private car. In fact, it is often the single most important action that consumers can take.

Milk and dairy products cause many of the same environmental issues as eating beef and lamb. You need ten litres of milk to make one kilogram of cheese. So, a climate friendly diet would not just be vegetarian, it would have to be mainly vegan. This is where it becomes more difficult for many of us, as that means reducing our consumption of dairy milk, cheese, cream, yoghurt, chocolate and even dairy ice-cream. Dairy free alternatives to most of these products are of course available. Eggs are also higher carbon than most plant-based foods.

Drinking dairy milk used to be a rarity, particularly as until the advent of refrigeration it was a difficult product to keep fresh. During the Second World War, the UK Government promoted milk as a good source of protein during rationing, and then schools gave free milk to every child at school. It is only recently that the adverse environmental impacts of intensive dairy production have become more apparent. There are alternatives to dairy milk, including milk made from rice, soya, coconut, almond, hazelnuts or oats. All create less than one third of the emissions of dairy milk; but rice and almonds do require a lot of water to grow. A study by the University of Oxford concluded that dairy milk had three times the greenhouse emissions of oat milk, required ten times more land and twenty times more water. Although the impact of dairy milk does vary depending on where and how it is produced, oat milk always has a lower environmental impact.

Milk from soya grown in North America or oats grown in northern Europe have the lowest carbon emissions, and both could be scaled up without significant adverse effects. Of course, taste and price are important factors in our dietary choices. Oat milk is more than twice the price of dairy, but if it became a popular choice the cost would tumble through economies of scale.

Is eating too much dairy products or meat bad for your health? The arguments continue to rage. The National Health Service recommends that we eat no more than 70 grams of red meat per day. Many cheeses and meats are high in saturated fats which can raise your blood cholesterol level, increase your risk of a heart attack, and there is a link between eating a lot of red and processed meat and bowel cancer. However, calcium found in milk, is healthy for young children and for strong bones, whilst meat is a good source of protein and essential amino acids. Calcium is also found in green leafy vegetables, dried peas and beans and in calcium-fortified products. Few plant-based foods are as good as meat on their own, but it is possible to obtain all the essential nutrients by eating a variety of plant-based foods, including tofu, beans, rice, lentils and seeds.

Given tooth decay and the effects of alcohol, our choice of drinks is important for our health as well as the environment. Producing spirits, wine and beer has a significant carbon and land-use impact with little nutritional benefit. Energy intensive distillation (evaporation) is used to make spirits, but the biggest impact is from growing the raw materials such as wheat or barley. Sales of £5 billion make whisky Scotland's most significant export, yet it is a luxury drink whose production causes 2% of Scotland's emissions – from agriculture, distilling, bottling and distribution.

Meanwhile, bottled water can cost as much as petrol, around £1 per litre compared to Scottish tap water at less than one penny. And the carbon emissions from bottled water can be around 2,000 times (300g:0.15g) per litre more than those of tap water with no health benefits. In fact, tap water is safer as it is treated and subject to more stringent testing.

This is all getting rather complicated to expect the average consumer to understand. Supermarkets should make these complex choices for us by promoting healthier and lower carbon products backed up with mandatory food labelling to help us to make good choices. The food company Quorn has promised to provide a carbon footprint for all the products it sells in the UK. They print the carbon footprint per serving or per kilogram on the packaging. An independent consultancy, the Carbon Trust, certify the calculation adding a degree of authenticity.   This information if widely available and understood could help consumers to make better choices.

****

The carbon impact of growing the food and drink that we consume is complicated and varies between seasons.   The impact on water, land-use and wildlife are a further complication.   However, we can make some generalisations which will help us to make more informed and sensible choices.

Fundamentally, it is best to grow food where the climate and soils are best suited to that crop.   So, bananas grow in the tropics, grapes in warm climates and apples prefer a cool and moist climate.   Planting in appropriate climates and soils reduces the amount of fertiliser, pesticide, artificial heat and irrigation required.   It is better if farmers grow crops using rain-fed water rather than crops that depend on irrigation. Large irrigation schemes, involving dams or pumping groundwater, can cause immense damage by diverting water from rivers and lakes, sometimes to the extent that they dry up destroying wildlife and fishermen's livelihoods.   The clear message is to buy food from regions that can grow it in the most naturally productive conditions (soil, water and climate).

Eating locally grown food does help to support farmers and local supply chains, is often fresher and therefore tastes better and it will reduce transport emissions.   But on average, transport is only 6% of the carbon footprint of what we eat, so the popular advice to eat local is frequently not the best advice, at least from a water-use or carbon perspective.

Due to the vast size and carrying capacity of ships, carbon emissions per tonne/km can be surprisingly low.   Apples and wine transported by container ship across the world are not as bad as you might think. Rail is five times more carbon intensive than shipping, lorries 20 times and freight carried by aeroplane over 100 times. So, depending on distances, the lorry journey to and from ports can be a significant element of the total transport emissions of goods carried across the globe. Avoid all foods that have been flown in. Even although this might provide jobs, for example highly perishable asparagus from Peru and green beans from Kenya, it is simply not a sustainable solution to improve the lives of people in low income countries. Avoid eating strawberries that retailers have flown in and wait to eat locally grown ones that will be fresher and tastier. Retailers should be forced to clearly label all foods that have been air freighted to help consumers make better choices.

Refrigeration is one element of the food chain that we do not normally consider. It keeps food fresh for longer which can reduce food waste. The entire food chain uses refrigeration from storage, processing, distribution, supermarket chillers to home fridges and freezers. But it is energy intensive, and it usually uses chemicals which can damage the ozone layer or add to global warming if they leak to the atmosphere. Interestingly, the carbon footprint of milk consumed in the UK is higher than in France. In the UK, most people buy refrigerated milk, whilst in France long-life milk is popular. It does not need to be kept cold during distribution or storage. There is enormous potential to improve the energy efficiency of the whole refrigeration process but, in the UK, the first step would be to remove the tax exemption from diesel used for refrigeration in trucks. Another simple change would be to ban open refrigeration units in supermarkets which continuously leak cold air. Supermarkets can add transparent sliding shelves, doors or lids; and even better, they should sell cold foods in a separate 'cold' room – I have seen this in Iceland of all places. Automatic doors separate this room from the rest of the supermarket. Once again, placing a proper price on energy use and from the emissions from refrigerants will incentivise the innovation and investment required to adopt the most efficient refrigeration technology.

Packaging is another element of food's carbon footprint. Packaging is used to attract consumers to buy a product, preserve food, prevent damage to products, reduce food waste and make food easier and safer to transport. Good design can reduce its impact. Lighter packaging reduces the energy required to transport goods. Nano-linings which enable us to squeeze out viscous liquids like tomato ketchup will reduce food waste. Regulations can encourage optimal packaging, such as a requirement that all food packaging be reusable, recyclable or compostable.

****

The common-sense solutions to reduce the carbon impact of farming were considered in the previous chapter, but the biggest change can be made by our dietary choices – shifting from meat and dairy based diets towards a predominately plant-based diet.

Like emissions from aviation, there are no technological solutions currently available to eliminate methane emitted from ruminants such as cattle and sheep - it is an inherent part of their biology. And as people become wealthier there is a strong tendency to eat more meat. Globally, particularly in fast growing countries such as China, the demand to eat meat and dairy is increasing. The logic is that if we are to reduce our emissions substantially then we need to change our dietary choices.

This need not be an overnight transition to becoming a vegan. There are lots of intermediary possibilities such as eating less meat, adopting a predominately plant-based diet, a healthy Mediterranean diet or becoming a vegetarian. And eating a vegan diet is not necessarily healthy or environmentally sustainable. Careful choices still need to be made over the choice of plant-based foods, considering issues such as the water used to grow avocados and almonds, and the damage from growing palm oil and soya beans in regions that were previously tropical forest.

At present, in most high income countries, meat eating is the expected normal behaviour perpetuated through our culture. Restaurants offer meat-dominated menus, often with a single vegetarian option (sometimes fish) and more rarely a vegan option. Similarly, at workplace events and conferences the default assumption has traditionally been that all attendees want to eat meat unless they make the effort to tell the organisers in advance that they are vegetarian.

Across the world, many people, mainly the young, are opting to be vegetarian or vegan. This could be for animal welfare, dietary or environmental reasons. Even one person making this change can have a disproportional impact. Parents who cook for the whole family may decide to offer plant-based dishes to everyone to save the hassle of cooking two different meals. A similar dynamic may occur at dinner parties. Restaurants and food outlets are also responding to changing consumer requirements. Gregg's the baker sells vegetarian sausage rolls. Burger King offers burgers without meat albeit they cannot label them vegan because of the risk of cross contamination during cooking.

Some countries, including Argentina, Australia and the USA, have meat even more deeply embedded in their culture with strong agricultural and political lobbies that support the meat and dairy industry. Openly encouraging vegetarianism can lead to hostility and a social backlash. In the USA there is a campaign for 'meat free Mondays' launched by Paul McCartney – perhaps as far as one dares to go within that culture. All I can say is that things do change over time, and the climate change imperative is clear that we should shift away from cattle, sheep and dairy based diets.

****

Governments can choose to be more interventionist. The UK levies a sugar tax on soft drinks dependent on the level of added sugar. This tax has had a negligible effect on choices made by consumers, but a significant impact on drink manufacturers. Over half acted to reduce the amount of sugar in their products to reduce or avoid paying the tax. Ironically, the resultant lower than expected tax receipts are a measure of the policy's success. The retail trade argued that the sugar tax would affect poorer sections of society, but in fact there has been little impact. Even better, the tax receipts support school sports programmes and healthy breakfast clubs to combat childhood obesity.

Governments should levy new taxes on products with a high environmental impact such as meat, dairy and palm oil; and unhealthy foods such as cakes, biscuits and sugary snacks. Extending the sugar tax to all sugar may be the easiest way to implement part of this. The tax receipts can be directed towards activities that benefit society such as supporting local markets, greengrocers, allotments, community run farms and boxes of vegetables delivered to your door.

We could reduce obesity and improve our dental and general health by eating more fruit, vegetables and unprocessed food. In general, unprocessed food such as oats, flour and apples are lower carbon than processed foods such as cakes and biscuits which require energy in their manufacture. Massive information and education campaigns are required to shift society's attitudes and help people to wean themselves away from meat and dairy towards a predominately plant-based and healthy diet. Finland is a good example of a nation that has changed its citizens' attitudes to diet and exercise with a sustained campaign over several decades. In the 1970s they had the world's highest mortality rate from heart attacks. They introduced legislation and community-based interventions to reduce smoking, salt and cholesterol intake. The government discouraged farmers from producing as much milk and encouraged people to eat more fruit. Farmers responded by growing fruits suitable to their Nordic climate. The results are impressive; the number of men dying from heart disease has fallen by 65% and life expectancy for both sexes has increased.

'Meat Free Mondays' has been a long running campaign, but I am suggesting going much further, perhaps 'Meat Mondays' or 'Meat Weekends'. We need to reverse what we have been brought up with and what is normal in our society – the idea of meat and two vegetables for every main meal. The 'nudge' affect could work well here. For example, event and conference organisers normally ask delegates in advance if they are a vegetarian or have other dietary needs - making it sound like choosing to be a vegetarian is being a bit difficult or awkward. We can reverse this philosophy by serving a plant-based meal as the default option, with attendees given the option to opt in to a vegetarian or a meal containing meat. Of course, we could easily go much further. At the Scottish environment business awards in Glasgow only vegetarian food was served.

Governments should launch ambitious strategies to encourage diets that are healthier and better for the environment. Public procurement for school dinners, hospitals and prisons should offer more plant-based options. Consumers should be offered advice, very few of whom are properly aware of the multiple environmental benefits of eating plant-based food, and even fewer are comfortable with how to make good quality plant-based dishes. We should encourage and subsidise cookery classes in every community to 'retrain' consumers to have the skills to select plant-based products and cook healthy meals.

We should promote healthy diets and low carbon diets together. As demand for plant-based products increases the price should come down.

To summarise, we do not all need to become vegan, but if we are concerned about our planet, we should curb our consumption of meat and dairy products. We should consider beef and mutton, preferably from rain-fed pasture, as a treat. Many of us would benefit from eating less food overall and we should be careful to avoid food waste – imagine all the hard work by farmers and carbon emissions wasted for food that we simply throw away. Government should compensate farmers for these new taxes on meat and dairy and launch a massive campaign to support the public to make better dietary choices that would reduce carbon emissions, reduce pollution and be good for our health.

Chapter 23:
Shopping and Buying Stuff

Capitalism seems to thrive on humans using energy, extracting more raw materials and consuming more and more products. Peer pressure at work, home and leisure backed by clever marketing entices us to buy more 'stuff' – we are bombarded with a constant stream of adverts and subliminal messages. The current levels of consumption in high income countries already cause environmental damage and are unsustainable.

Of my five common-sense principles, the most relevant to shopping is to 'consume carefully'; but 'avoid waste', 'price carbon pollution' and 'nurture nature' are all relevant too.

Before buying any goods or services, we should ask ourselves if we really need it, how it was manufactured, will it make our lives better and how will we dispose of it. And buying stuff is not just about its impact on the climate and nature. It also impacts on your personal finances. Some are lucky enough to be able to buy whatever they want, whenever they want. But for many, spending too much gets them into debt. If you buy a large house, perhaps larger than you need, or invest in a house extension, then you might have difficulty paying your rent or mortgage. If you buy new kitchen units, a new car every three or four years, excessive Christmas presents, lots of unnecessary new clothes or even multiple take-away coffees then these decisions will impact on your short-term finances and potentially the long-term. It is likely to delay your ability to retire with a comfortable lifestyle.

The Office for National Statistics (Material Footprint in the UK) estimates that the UK used 1.2 billion tonnes of materials in 2017, averaging over 18 tonnes per person. This includes timber, metal ore, construction material and fossil fuel. Remarkably, we import 80% of this material, mostly from China and low income countries, and this proportion has increased over recent years. Despite being the country that led the industrial revolution, and with a landscape scarred by thousands of abandoned mines, we import nearly all the metals we now consume. More than anything else, these statistics demonstrate how high income countries have outsourced the mining, harvest and extraction of raw materials that they need to feed their consumer lifestyles. Other countries feel the environmental impact of our excessive consumption most acutely - out of our sight, out of mind. And, the environmental damage overseas is likely to be greater due to less efficient production processes and transport. Therefore, we need to reduce consumption and improve resource efficiency and recycling.

If you want to reduce your consumption you need to understand the hidden world of product lifecycle footprints and how and where manufacturers source their raw materials and make products. To cut your use of water, many of us would consider buying water efficient appliances such as a new dishwasher and washing machine or avoid running the tap as we clean our teeth. But we could make a far bigger impact by changing our consumer shopping habits. It can take 17,000 litres of water to make one kilogram of chocolate, 15,000 for one kilogram of beef, 2,700 to make one cotton t-shirt and 1,000 for one litre of milk. By comparison, soya milk only needs 300 litres. But not all water use is equally harmful. Rain-fed crops cause less harmful impact than those that rely on irrigation in semi-arid regions.

All products have an impact on the environment, usually an adverse impact. To consider some everyday items: wood is used to make toilet paper, as are 60 billion pairs of disposable chopsticks used each year. Wet wipes contain plastic which can block sewage pipes if flushed down a toilet and birth control pills release chemicals into rivers. In the UK we dispose of 3 billion nappies containing wood pulp and plastic, mostly to landfill each year. In Japan, with its elderly population, more nappies are sold for adult use than for babies. Oil is used to manufacture these nappies, one trillion disposable bags and to make polystyrene packaging which is rarely recycled. We dispose of two billion razors containing plastic and metal to landfill each year, and chemicals and heavy metals are mined to manufacture electronic items and batteries.

Other chapters in this book consider food, transport and tourism. This leaves clothing, followed by electronics, household products and personal care as the consumer categories with the highest environmental impact. British consumers buy around one million tonnes of new clothes per year. The fashion industry has access to international supply chains and has successfully, some might say ruthlessly, exploited this to source cheap raw materials and garments, often made in factories with poor employment rights and low environmental standards. The result is that consumers have access to extremely cheap clothing. The problem is not clothing, it is fast fashion. Much of what we buy is cheap, creates pollution, is made from poor quality materials and workmanship and does not last long.

The carbon emissions to manufacture and dispose of all these clothes are more than that from international flights and shipping combined. In fact, textiles and clothing are the fourth highest carbon emitting category of consumption in the EU after housing, transport and food. 97% of clothes are made from virgin fibres – mainly synthetic (plastic) fibre or cotton. From a carbon perspective, clothes made from synthetic fibres are likely to have a lower footprint than from cotton or wool. However, all clothing has additional and complex environmental impacts whether made from natural materials such as wool, cotton and leather or synthetic materials such as polyester and nylon. These include fertiliser and water to grow cotton, oil for synthetic clothes, chemicals to process fibres, water pollution from dyes, and pollution from transport and shipping. New materials like Tencel, made from wood pulp may be a more sustainable choice.

After the consumer has bought their clothes, there is the recently highlighted issue of the release of plastic microfibres when washing them. Then 70% of clothes are discarded to landfill, and many that are recycled are only made into lower value products such as cloths – it is difficult to recycle clothes that often contain mixed materials. For example, spandex, sold as Lycra, is a synthetic material with elastic properties that can be blended into other fabrics. Existing technology cannot separate it for reuse or recycling.

The carbon impact of clothes is more than just from manufacturing and retail. Consumers use even more energy at home to wash and dry their clothes. Proctor and Gamble calculated that 90% of the lifecycle carbon footprint of their washing powder arises when their customers wash clothes at home. This spurred them to develop a new powder that works equally well at lower water temperatures – you can now wash clothes at 30^{0}C, saving energy at home. Of course, wearing clothes for longer before washing them will save even more. Mondays used to be washing day; now some families have their washing machine on every day, with every wash using water, energy and releasing thousands of plastic microfibres into our rivers. The best solution is for consumers to be willing to pay more for higher quality and durable clothes, and then to keep and wear clothes for longer. The retailer can supplement this by offering long guarantees, backed up with repair and refurbishment services. GRN Sportswear sells quality, sustainable sportswear for teams. Most of their raw materials are synthetic, from recycled sources or off-cuts. They avoid using mixed fibres, design their products to be durable and are exploring the potential to chemically recycle used clothes to become the raw material for new garments.

Another option for the fashion conscious - those who feel the need to wear different clothes when they go out to events - is to rent outfits from companies like H&M Group which you return to be cleaned and re-used. As a final thought on clothing; it should be socially unacceptable for people, newspapers and social media to criticise anyone, friends or celebrities, for wearing the same clothes twice in public.

As societies become wealthier the consumption and associated carbon footprint of cleaning, personal care and hygiene products increases, driven by marketing that persuades us to buy products to make our homes and ourselves look or smell attractive. Daily or more frequent showers are a recent phenomenon. Over-use of cleaning products may adversely affect our health - after all they contain a cocktail of chemicals designed to destroy living bacteria. Products are often over-packaged or in single-use packaging. Single-use razors, toothpaste tubes, nappies and feminine hygiene products create vast quantities of waste. There are alternatives; quality razors, toothpaste tablets, washable cloth nappies and menstrual cups; yet our society, driven by marketing pushes us towards convenience rather than choices that are potentially less toxic to our bodies, cheaper and better for the environment.

Bottled drinking water is another product with a high carbon footprint – around 2,000 times more than tap water. Yet, bottled water has become popular, initially in countries where there were doubts over the quality of the tap water, but clever marketing has successfully spread this 'lifestyle choice' to countries where the water quality is good. So, people become accustomed to buying a product in a single use plastic bottle that is as expensive as petrol. You can even buy water shipped across the world from the remote island of Fiji in the Pacific Ocean. The economics that enable this trade are crazy - the price of fossil fuel used to manufacture the plastic bottle and to ship this heavy product across the world does not take carbon emissions into account. In the absence of the sensible economics that we need, retailers have a responsibility in the choices they offer, and consumers in the choices that we make.

****

My belief is that we can enjoy fulfilled lives without consuming as much, particularly buying new physical goods. But governments need to set the framework to shift our consumerist society from one that mines and extracts raw materials from the earth with impunity to one that minimises the need to extract virgin raw materials. Businesses should manufacture long-lasting goods that are repairable, retailers and salespeople should focus on durability, and consumers should demand long lasting quality.

Governments should shift the burden of taxes from taxing our income to taxing the extraction of raw materials, land ownership and our expenditure on goods. The aim is not to raise the overall amount of tax collected, but to redirect our taxes to benefit both society and the environment. By raising tax in one area of the economy, it will be possible to reduce another tax elsewhere or to provide directed subsidies and assistance to those most in need.

Value Added Tax (VAT) is a good example of a tax on our consumption, although governments cannot raise it too high without people trying to avoid paying the tax and creating a black market. But in the UK many products are exempt from the full rate of VAT such as gas, electricity and building new houses. All products should have VAT levied, unless they have a specific environmental benefit or strong social purpose - insulation, heat pumps and perhaps electric cars. VAT on food in the UK is complicated, but in general, it is levied on food served in restaurants (as a service) but not food bought in a supermarket. As a minimum, all unhealthy foods and those that have a high carbon impact should have VAT levied – including sugar, rice, meat and dairy products. It might be better to go further, and levy VAT on all food except those with specific health and environmental benefits such as fruit and vegetables.

Governments should tax or place a levy on new clothing. As all clothing has an environmental impact - methane from sheep growing wool; fertiliser and water use for cotton, oil use and microfibre pollution for synthetic clothes - it would be much simpler to levy one rate of tax across all new clothing. Some of the tax could be used to finance reuse and recycling schemes.

Peer pressure and 'keeping up with the neighbours' is a hidden, but powerful force that drives our desire to buy new goods. Peer pressure only works through the support of the population, or at least their tacit support, often reinforced by marketing messages. On-line bloggers or 'influencers' are paid to promote short-life fashion items. As such, if the opinion of key influencers and subsequently the majority were to shift, then peer pressure could be reversed away from continuous consumption towards frugal consumption, or what I'd prefer to describe as 'quality' consumption. Our expenditure could also be redirected towards goods that have a clear long-term benefit such as investing in triple glazing and solar panels for your house rather than spending, on say, a kitchen makeover or a luxurious holiday.

For example, changes in public attitudes could accelerate the shift to electric cars. The 'dieselgate' scandal, where Volkswagen cheated on diesel emission tests, raised public and regulators' consciousness of the health impact of pollution from burning diesel. Burning petrol is environmentally harmful too, contributing to climate change and air pollution. Environmental warning labels should be placed on petrol and diesel pumps, like the health warnings on packets of cigarettes. This would reinforce the message amongst motorists that burning fossil fuels harms other people and nudge motorists to think more about the benefits of buying an electric car. Taking a long-term view, I am sure our society will look back and wonder why people thought it was ever acceptable to drive private cars which emitted harmful gases within our cities.

LEGO is an example of a company that has successfully built up its global business through the power of marketing. LEGO has manufactured 400 billion plastic bricks since 1958. Children and adults use the standard bricks to build anything within their imagination and they can be reused repeatedly - very few are thrown out. However, once the market for standard bricks became saturated, LEGO began to design specialist pieces for specific models. Once the consumer loses the instructions, these specialist pieces are not as useful and are more likely to be thrown out to landfill.

We need to use all the tricks that marketing people have developed and perfected over the last hundred years to successfully persuade us to buy more, bigger, faster, newer, fashionable - but to reverse this to buy less, better quality, compact and functional, useful and meaningful, and long lasting. Capitalism can thrive under this new paradigm if consumers are willing to pay more in return for better quality goods or enter into a contractual agreement with retailers to purchase ongoing services.

Business should lead this 'U' turn, supported by consumer education. The government should require retailers to print more information on product labels, apply selected taxes to shift our behaviour and introduce new regulations to reduce unnecessary consumption (such as a ban on plastic straws). The media, social-media influencers and popular culture can support change too. Imagine if a well-known family on a soap opera such as EastEnders were to repair their clothes or invest in solar panels or triple glazing. This might have a bigger impact on our behaviour than a government led campaign.

Business will respond to this consumer pressure by engaging with their supply chain and offering better quality products to consumers. Some businesses rent or lease products rather than sell them outright. An example is new tyres for cars. The garage offers a wide choice made by different manufacturers and of varying quality, but the consumer is wary of fully trusting the salesperson. All you want is the best value tyres that are long-lasting, safe and a reasonable price. Then, the garage charges extra to take away your old tyres for safe disposal. A better model would be for the garage to rent tyres to you for a monthly fee based on your estimated mileage, possibly linked to a tracker to monitor your driving style. The manufacturer and garage now have an incentive to sell the best quality tyres that will last longer, be safer and emit less micro-plastic.

****

If the carrot is voluntary initiatives to support better environmental performance of products, society would also benefit from using a stick to ban some products and curtail the marketing of others. Sometimes the aim would be to reduce the carbon impact of products, such as incandescent lightbulbs, but often it is for other good reasons such as reducing litter or to set a visible example to consumers of more desirable behaviour.

Society has already introduced restrictions on marketing of cigarettes, alcohol, speeding cars, gambling and fast food. Why not extend this to other harmful products such as gas guzzling cars, long-haul flights, beef and dairy and even bottled water? Controversial maybe, but marketing should be a force for good, focusing its efforts on products that are beneficial to society.

The EU is planning to ban certain single use plastics including cutlery, plates, straws, balloon sticks and cotton buds. Other products containing plastic that could be banned include cigarette filters, microbeads in cosmetics and toothpaste, grocery bags, coffee pod capsules, tea bags, glitter..... the list is endless.

Other products that do not degrade easily in the environment include latex balloons, floating Chinese lanterns, labels stuck on fruit, metal staples on some tea bags and most wet wipes. And, from personal experience from my compost heap, manufacturers should redesign Christmas crackers to degrade more easily.

Other products of questionable value to society include disposable razors, outdoor patio heaters, disposable barbeques, metal foil helium balloons, Styrofoam cups, bottled water, electric toothbrushes, artificial grass, polystyrene food containers, incandescent light bulbs, oxybenzone in suntan lotions (toxic to coral) and unrecyclable pens and other stationery.

Some argue that the problem is not plastic. It is the inappropriate disposal of products such as people flushing used cotton buds down the toilet. There is some truth in this, as most plastic pollution in the oceans comes from fast developing and low income countries without waste collection and disposal facilities. However, a ban on these products sends a powerful message to consumers to discourage frivolous consumption and to business to promote eco-design and the circular economy.

****

It is not just the carbon and climate impact of buying stuff that we should consider. There are lots of other issues that arise from manufacturing products such as excessive use of water, impacts on land-use, mining and the rights of workers. There are numerous business and charity led initiatives to help guide us towards making better choices, some more useful than others.

Fairtrade encourages farmers in low income countries to be paid a fair wage and to work in safe working conditions. The Marine Stewardship Council certifies fish based on science-based requirements for sustainable fishing. The Roundtable on Sustainable Palm Oil has developed environmental and social standards for producing and processing palm oil. The Alliance for Water Stewardship encourages companies to look after water wisely and to recognise the social, cultural, environmental and economic value of freshwater. The Rainforest Alliance aims to create a better future for people and nature by improving the livelihoods of farmers and forest communities. Business should engage with these schemes and demand the tightest controls possible, whilst consumers should consider buying products which demonstrate compliance with these standards. Pressure groups should continue to monitor these schemes to ensure they maintain high standards.

All these initiatives have their critics, but we should not discount them for not being perfect. Some are independent, whilst others are led by industry with the underlying objective to stave off tighter regulations. The best schemes are independently managed but also involve meaningful discussion with consumers, pressure groups and all the players in the industry.

The Roundtable on Sustainable Palm Oil has certainly been subject to some valid criticism. Iceland, the frozen foods retailer, even decided to phase out all palm oil from its products. Only 20% of all production is within the Roundtable scheme, and it has not yet succeeded in one of its objectives to prevent farmers and corporations from establishing new palm oil plantations in virgin rainforest. Although boycotts are good for making a point, they then tend to exclude that source of pressure from future negotiations, reducing the influence and pressure on the industry to go further.

Overall, it is good for consumers to engage with all these initiatives, to buy labels that they trust and to continue to pressure campaign groups and business organisations to improve the sustainability of these complex international supply chains.

****

Another way of potentially reducing emissions from our consumption of physical goods is to 'go digital' and consume services rather than manufactured physical goods that you can touch. We can read books on a screen, the music industry has rapidly moved from compact discs to on-line music, and there is now little need to print photographs. Videoconferencing can remove the need to travel, virtual models and testing can be used to perfect new designs and products, and some training can be carried out on-line or even on a simulator rather than in real life. These all have the potential to reduce our consumption of resources, but as always it is not straightforward with the outcome dependent on the resulting behaviour change. There is no golden rule, but a book read several times by different people then recycled at the end of its life may have less of a carbon impact than buying an e-reader and not reading many books on it. However, if you already own an e-reader then it will be better to use it.

In future, technology will enable more individualistic consumption such as diets designed for individuals, drugs and dosages dependent on our DNA and clothes designed to our individual body shape. Technology can enable ever more consumption or can be used to enable consumers to make more ethical and lower carbon choices. It can be a force for environmental destruction, or we can direct it to tackle societal and environmental problems. Our choices can make a difference.

All digital content, including emails and web searches, require energy. Your computer, tablet or phone use electricity at home; but of greater concern is the electricity used by servers and data storage. Power is used to access a remote data centre every time you download or stream from the internet. This is all hidden from us, but with the growth in computing power and storage capacity it is rapidly becoming a major issue. And, in future, power consumption will increase with access to high definition videoconference facilities offering realistic handshakes, immersive virtual tourism experiences and 'mining' digital currencies such as Bitcoin.

Miniaturisation, such as moving from desktop computers to laptops to mobile phones, has reduced power consumption. Smart phones also have the potential to replace several devices with one small device that has an overall smaller carbon footprint. This could result in a significant reduction in your need to buy other electronic devices including a calculator, watch, music player, games platform, fixed line telephone and camera. But if you buy these products in addition to the facilities on your phone then there will be no saving. Your smart phone can also enable you to work, bank and shop from home, and it provides information and access to education which can be particularly beneficial to people in low income countries with poor access to physical services and infrastructure.

Overall, it seems that digitisation has enabled a massive increase in consumption which is offsetting much of the potential savings from digitisation. People now take hundreds of photographs stored on the cloud, but still print some. We now have access to thousands of songs on-line rather than the dozens most of us owned before when we bought compact discs. At work there has been much talk of the 'paperless' office. Whilst this is theoretically possible, many of us now access far more emails and download many large documents, some of which we print for our convenience. As always, paper reduction only occurs if behaviour changes along with the new technology.

Despite some misgivings, if we use technology to our advantage, and if renewable sources generate the electricity then a shift to a digital economy will reduce our emissions. But, for those of us in wealthy countries, there is little evidence that buying more stuff and buying poor quality disposable items makes us happier. We need governments to shift taxation from income to the extraction of virgin raw materials and to consumption, to require clearer labels on products and for more regulations to support the circular economy. Companies should engage with their international supply chains to raise environmental standards, to continue to innovate and invest in resource efficiency and to provide products that offer quality and long life - possibly at a higher price. We will lose some jobs in production and manufacturing, but create other jobs in services, maintenance and repair. Consumers should use their peer pressure to influence others to buy beneficial products and to stop the existing pressure to continuously upgrade and buy new. We need all of this, and more, to prevent the current global inequity of the lifestyle of affluent people continuing to rely on exploiting the raw materials, nature and wildlife in poorer countries.

Chapter 24:
Business and Industry

We can trace most of our carbon impact and environmental destruction to the food, goods, services, buildings and infrastructure that industry and business provide. One hundred companies directly account for 70% of global carbon emissions. Consequently, you could argue that it is up to business to 'solve' climate change, but this is too simplistic as businesses do not exist in a vacuum. They provide goods and services that consumers want to buy, and they operate within a framework of regulations set by governments.

Despite claims by some environmentalists, we do not have all the technology needed to tackle emissions from industry - certainly not cost-effective solutions that industry is likely to adopt soon. An effective price on carbon emissions would speed the transformation of industry that we need. However, a carbon tax would not be enough on its own as industry might absorb the tax and put up the price of what it sells. We need a strong industrial policy and international cooperation, backed by a carbon price to tackle the emissions from industry. To transform the sector, we need targeted subsidies to build factories to trial and adopt new technological innovations. Of my five common-sense principles, 'embrace efficiency, avoid waste' and 'nurture nature' are the most relevant to business. Adopting a long-term approach to investment would help too.

****

It is important to realise that there are different ways to structure a business. There are multi-nationals, public limited companies (PLC's), private limited companies, social enterprises and sole traders. Some, such as family owned businesses, have more control and influence over what they can do than others. PLC's have a legal duty to maximise value for their shareholders, so to reform this and drive change will require new regulation. Some businesses have access to their own funds; others are indebted to banks or shareholders and are tightly controlled by investment institutions. This means that some companies have freedom to act and can change quickly. Others are susceptible to positive or negative pressure from consumers, investors and campaign groups. What unites all businesses is that they must make money to survive in the long term.

In the UK, our society allows public and private limited companies to legally operate with 'limited liability'. The company can then take risks to make money without the risk of the directors losing their personal assets including their home if it goes bankrupt. We could alter this balance to mandate limited companies to 'do good for society', including environmental objectives, in return for continuing to benefit from this legal privilege of limited liability. Companies such as cooperatives and social enterprises already do this, and a significant minority of other businesses already support environmental and social causes even if it is not directly profitable. We need all companies to work to create environmental and sustainable value for our society. All companies should be socially useful.

There are those that claim that business is not interested in the environment, only in profit. That business will only invest in environmental improvements to save money, to sell more of a product or if regulations, their investors or customers push them to do so. But most businesses need to constantly change to remain viable. Twenty years ago, most could not see a need to have a web presence or to sell on-line. However, the world soon left behind any business that did not move with this change. Similarly, as climate change and environmental issues enter the public consciousness, companies will need to offer sustainable choices to consumers. Businesses can differentiate themselves from their competitors to gain competitive advantage by becoming more sustainable. The benefits can include cost savings, enhanced reputation, increased revenue from selling new products, gaining the support from their local community, compliance with legislation, and importantly an increase in employee motivation and staff retention.

****

A sustainable business model is one where business can successfully align the profit motive with environmental benefit. The opposite is where they compete. In the same way that it is difficult for a tobacco company to promote health, it is inherently difficult for an oil company to be truly sustainable as their profits are fundamentally dependent on locating, extracting and selling ever more oil. Even so, they have no excuse not to be as carbon efficient as possible, for example, to avoid flaring 'waste' gas.

Sustainable business models closely align to the three circular economy principles of design out waste, keep products and services in use and regenerate natural systems; supported by the concepts of reuse, repair, remanufacture and recycle. Encouraging reuse, repair and remanufacture is all down to good and sensible product design, perhaps using fewer composite parts, standard components and avoiding parts being 'welded in' and becoming inaccessible to dismantle. To encourage a circular economy, we need stronger regulations on 'fit for purpose' consumer rights, improved energy efficiency of new products and to implement the EU's Eco-Design Directive, or equivalent. Government should tighten and enforce all of these. We should move harder and faster - for example, government should regulate for compulsory long guarantees on electronic goods, furniture, cars and machinery to encourage manufacturers to design products to last longer and be more easily repaired. This single measure might have a transformational effect. Manufacturers could no longer design products with a short-expected lifespan or with built in obsolescence; instead they would design higher quality products to maintain a good relationship with the consumer over the long lifespan of the product guarantee. Consumers may be willing to pay more in return for higher quality.

Many of us have had the experience of buying an electrical product like a washing machine only to find that a fault develops soon after the guarantee expires. The appliance is difficult and expensive to repair, either because parts are welded, or it is difficult to source spare parts. An alternative model would be to lease washing machines. The retailer is then responsible to repair, refurbish and possibly to upgrade it. This model ensures that the manufacturer and retailer both have an incentive to manufacture and sell quality products. The French government even passed a law in 2015 to ban planned obsolescence – techniques that deliberately reduce the life of a product to increase the volume sold.

To boost remanufacturing and recycling, an extended producers' responsibility would force manufacturers to think about how their products can be safely and responsibly disposed of at the end of their useful life. This is not something you hear people campaign about, but will force manufacturers and retailers to take responsibility for the products that they sell. For example, retailers should be obliged to collect old bed mattresses in the same vehicle used to deliver new ones. By setting up such an efficient collection system, it is possible to refurbish or recycle old mattresses and avoid the need for consumers having to pay for them to be taken to landfill. Similarly, retailers should take back items such as old running shoes and tennis rackets when people buy new. Standards can also be set for new products to require an increasing level of recycled content over time. This will create a market for recycled goods and therefore help to drive the collection of discarded goods. More comprehensive and clear labels on products would make it easier for consumers and recyclers to recycle goods. Reducing the number of plastic polymers in packaging and clothing would make it easier to segregate and identify what is suitable to recycle. Successful recycling across the economy will reduce the volume and cost of importing virgin raw materials.

Optimising packaging is a balance between avoiding the waste of over packaging with the product waste that can arise from poorly designed packaging. Amcor, who make packaging for consumer products, has pledged that all its packaging will be recyclable, reusable or compostable by 2025. It will be difficult for governments to regulate everything, but it should simply be unacceptable to manufacture and sell a product in packaging that is not easy to compost or recycle such as crisp packets made from a hard to recycle metallised plastic film or from aluminium foil within multiple layers. Manufacturers, retailers and consumers should ask questions about the products and services that they make, sell or buy. We should all reconsider our role if we produce, sell or buy products or packaging that will end up in a landfill site within days of its purchase. Pressure from consumers can influence businesses to offer us better choices. This needs to be backed up by good collection facilities to recycle unavoidable packaging.

Companies can sell consumer products in recyclable and refillable containers. Proctor and Gamble has trialled this with washing powders, dry foods and even ice cream and deodorants. The container can remain the property of the retailer; the consumer has an incentive to return it - often a discount off their next purchase. To reduce the cost and carbon emissions of transport, items such as cleaning liquids, which are 90% water, can be sold and delivered to the consumer who adds water to the concentrated cleaner at home. We could go much further and introduce a standard set of common containers for trade and consumer products. Manufacturers and retailers could print their own labels on these containers. After use, they could be collected, washed and redistributed to manufacturers within the scheme.

On an international stage, 500 companies have signed up to the New Plastics Economy Global Commitment, facilitated by the Ellen MacArthur Foundation. They have pledged to eliminate unnecessary plastic by 2025, invest in innovation and circulate used plastic within the economy to keep it out of the environment.

Companies can also embed circular economy principles into their business strategies and marketing. Renault, the car and mechanical plant manufacturer was an early adopter of the circular economy in 2000. It set up a factory to remove and remanufacture old engines and gearboxes for resale with the same warranty as new products. This saves energy and the purchase of raw materials. They also strip down badly worn parts to recover copper, steel, aluminium and plastic. Patagonia, the outdoor clothing manufacturer and retailer, famously launched a marketing campaign entitled "don't buy this jacket" pointing out the full environmental impact of their products - their honesty and integrity resulted in an increase in sales.

Offering products as a service is one strategy that can increase customer engagement and boost sales. Hewlett Packard sells an 'instant ink' service. They remotely monitor the level of ink in printers and send new ink at a low price when required. The customer posts empty ink cartridges back for re-use. It is this sort of thinking that is required, as good 'reverse logistics' are often needed to enable packaging to be reused - it is easier to reuse, refurbish or refill a product if the seller retains some form of contact with the buyer. Signify offer 'lighting as a service' where they install energy efficient light bulbs in an office or factory and retain ownership of the bulbs. This incentivises them to sell quality long lasting bulbs, and they can introduce the latest energy efficient technology if this proves to be a cost-effective investment. Rolls Royce offers a 'power by the hour' service for its aeroplane engines with remote monitoring by engineers which predicts when they need to be serviced, saving cost and increasing safety.

Even more remarkably, the software industry can now send updates and improve its software to millions of consumers instantaneously over the internet. Tesla can now send upgrades to its customers' electric car engine management system to fix bugs and even to enhance and improve battery performance. Companies can improve products that we have bought over time, rather than allow them to deteriorate and depreciate.

****

For many years I have been a judge for the Scottish Business Environment Awards (VIBES). This involves visits to companies to learn about a new environmentally friendly product they have developed or how they have reduced their consumption of materials, energy or water. These visits can be inspiring with case studies such as CMS Enviro available on their website. They manufacture and install double and triple glazed windows in social housing to reduce energy use and fuel poverty. They have invested in energy efficiency for their factory. Their product is inherently beneficial for the environment, but in addition they collect and recycle the old windows that they replace. They restored and upgraded a building to become an academy to train installers and to educate the wider public including school groups. This comprehensive approach is inspiring – a socially useful product, designed well and integrated with training staff and collaborating with their sub-contractors.

From my experience, the best sustainable companies demonstrate several common attributes. Firstly, there is a belief and drive from the managing director or board. There seems to be a shared mind-set between a managing director that cares about their employees and one that cares about their impact on the environment.   Next, there is the understanding that you cannot enhance the environment on your own; you must partner with your suppliers and sub-contractors and maintain an on-going relationship with your customers.

The best companies will undertake a wide range of sustainable initiatives and activities encompassing everything that they do. They engage with their employees, invest in staff training and integrate environmental considerations into their strategy, purchasing, production, marketing and finance. They measure their environmental impact and set ambitious targets. They consider the implications on the environment of all their investments, and specifically invest in energy and resource efficiency, on-site renewable energy and to develop more environmentally friendly products to sell.   They collaborate with suppliers, and partner with community and social groups. They build brand loyalty with customers by maintaining an ongoing relationship with them.   They invest in innovation, working with universities or other research institutions, with sustainable design, and circular economy principles at the heart of all new product development. They might re-focus to provide a service, rather than sell a one-off product.   For example, Interface, the carpet tile manufacturer, offers to rent carpets and tiles to its business customers. This provides the financial incentive to optimise the durability of the carpets that they sell and to ensure that they are easy to refurbish or recycle at the end of their useful life.

What is interesting is that when you visit a company that really believes in sustainability, you find that they are never finished. They always have several sustainability projects on the go, with more ideas and further plans. The end goal is not to cut their environmental harm by 'x' percent. The goal is to actively benefit and improve the environment which is a never-ending mission.

Companies can demonstrate their sustainability credentials by obtaining formal certification. There are several schemes including ISO 14001 and B-Corp. Patagonia is an example of a company who have gained the prestigious B-Corp certification. They focus on quality and improving the design and durability of their clothing and other products. They carefully consider how they source their raw materials, using natural materials where possible and offer long product guarantees. They also offer a clothing repair service, or customers can trade-in products to be refurbished or recycled in return for a discount off new purchases. Yorkshire Tea has been certified as carbon neutral across its operations, supply chain and products. They installed solar panels at their offices in England, but more interestingly invest to improve the energy efficiency of tea packaging factories in Kenya, they plant fruit and nut trees to sequester carbon around Kenyan tea plantations and distribute fuel-efficient stoves in Malawi.

However, we need to guard against 'greenwashing', where a business makes environmental claims that are dubious, wrong or offset by other environmental damage they cause. Many multi-nationals and oil companies are prone to this. Given what we now know about climate change, what really counts for companies in the fossil fuel sector is how robust their medium-term plans are to diversify into the low carbon economy, backed up by hard cash such as the proportion of their innovation and investment in renewable energy.

As a first but necessary step, businesses should calculate their carbon footprint. If you do not measure it, it is difficult to know how to even begin to improve. In the UK, recent legislation requires businesses with over 250 employees to calculate and publish the carbon footprint from their direct operations – the electricity and gas to run their factories, business travel and distribution. Government should extend this requirement to all but the smallest of businesses.

Of course, the impact of business is not just what a company does directly. It includes the impact from their suppliers, from those who consume the products they sell (e.g. energy in use) and from the ultimate disposal of the product. International supply chains can be complicated and murky but failure to act responsibly can lead to a consumer backlash. This applies to commodities such as tea, coffee, palm oil and cotton; and to manufactured goods such as t-shirts. Businesses should audit their supply chains to ensure that they are ethical and meet environmental standards.

Palm oil, with its high saturated fat content, is an ingredient in half of the products sold in our supermarkets. It is in food such as bread and chocolate, cosmetics like shampoo and lipstick, and is blended into transport fuel in south-east Asia. It is a productive crop and is cheap because the adverse environmental impact of palm oil plantations is not factored into the price – palm oil drives the destruction of primary forests in south-east Asia, destroying areas of high biodiversity value including the habitat of orang-utans. Some companies are members of the 'Roundtable on Sustainable Palm Oil' although even this group has its detractors. Although its members promise not to clear new forest, the damage has already been done, or is displaced to other areas managed by companies who are not members. This is a complex area, but one which large multi-national companies can grapple with and influence.

Many businesses, particularly chains and multinationals, have set targets to reduce their environmental impact. The best targets are medium-term and ambitious and are most likely to be achieved if there is a link between directors' pay and achievement of the targets. Some targets help to drive progress, others less so. For example, long-term targets become lost in the noise of general business activity as companies open new branches, increase their turnover, merge or takeover other businesses. It is more helpful to set short and medium-term targets that result in significant investment funds being allocated to achieve them.

Big businesses now seem to be competing against each other in their level of carbon ambition, publishing ever more ambitious carbon strategies and carbon reduction targets. The devil is in the detail however; whether they have any real intention, backed up by funds to innovate and invest, to achieve what they announce.

Traditionally companies have set targets to reduce carbon emissions from a pre-determined baseline, for example to reduce carbon emissions by 'x%' by 2025. Some set targets that allow for growth in the business, for example, reducing carbon emissions per pound of turnover by 20% from 2020 to 2025. Such targets can be easily met if there is high inflation, but they do provide flexibility - for a fast-growing business, targets that exclude any reference to growth in turnover are more difficult to achieve and are often quietly dropped following any significant change in the business structure. A better way to set targets is a percentage reduction per unit of production as this provides a suitable adjustment for any growth in the company but still requires action to reduce emissions. The focus of all these initiatives is usually energy efficiency and sometimes investment in producing renewable energy. None of these targets require the company to transform, instead they promote incremental change.

Some companies have gone further, promising carbon neutral or net zero emissions. This is usually from their direct emissions as these are easiest to measure, control and implement. Net zero means that the company will offset any remaining emissions, usually by paying into an offset scheme. Companies as diverse as BskyB, Dell, Google, HSBC and Nike have committed to make their entire operations carbon neutral. Orsted, the Danish energy company has announced a net zero target for 2025. It has transitioned from a state-owned oil and gas company to focus on offshore wind with a vision for the world to be run entirely on green energy. Corporate Knights named Orsted in 2020 as the most sustainable company in the world. BP announced a net zero target for 2050 covering its operations, exploration and production, and to reduce the carbon impact of the products it sells by 50%. Currently its global operations and products account for a larger carbon impact than from the entire UK economy. 2050 targets are too remote to help with the climate crisis, but BP has also set a target to reduce oil and gas production by 40% by 2030 and to substantially invest in renewable energy. Encouragingly they announced that 36,000 staff will have their performance bonus linked to their success in meeting environmental targets.

Others have extended net zero to include their supply chains or even promised to offset all their historical emissions. Unilever announced a net zero target including its supply chain by 2039. Microsoft announced targets to achieve 100% renewable power by 2025, to be carbon neutral by 2030 and most remarkably, by 2050, to offset all its historical emissions since it was founded in 1975.

Can companies shift even further, away from inevitably causing environmental harm? It would be truly inspirational to transform targets from reducing the level of harm, towards a story of business opportunity, a chance to restore nature and to make a positive contribution to the environment. Net positive, or regenerative, means that the company promises to do this. Companies can achieve this by planting more trees than they cut down, producing more renewable energy than they consume or returning water to the environment cleaner than it was extracted. Companies like Patagonia, Unilever, Ikea and Natura certainly think that it is the duty of business to make money and to restore the environment. Unilever's Sustainable Living Plan aims to improve the lives of customers who consume their products through improved hygiene, health and well-being. They aim to be carbon positive across their operations by 2030 by generating more renewable energy than they consume and avoiding the need to pay for any offsets. Similarly, Dutch multi-national DSM sells products for health, nutrition and materials. They have targets to reduce their direct emissions by 30% and emissions from their supply chain by 28% per tonne of product manufactured by 2030. They already claim that two-thirds of their products are measurably better in terms of environmental or social benefit than their competitors.

****

Because of risk, uncertainty and interest rates, investors and companies often take a short-term approach to business investment. Company directors are often paid bonuses based on short-term financial performance. Long-term targets and clear policy, an area which governments tend to be poor at, could help to reduce the uncertainty and risk and would encourage business to invest for the long-term. Constant changes to energy policy, taxes and subsidies do not help.

Even worse, senior managers often consider investment in energy and resource efficiency as less ambitious than say investing in a new manufacturing plant to increase sales. This means that companies do not take advantage of all cost-effective opportunities to significantly reduce environmental impact. The EU's Energy Efficiency Directive requires large EU companies to assess and identify cost effective measures to reduce their energy use. We could go far further and extend this to a wider range of companies and to actually require companies to implement the measures identified over a number of years - supported by subsidies such as interest free loans. By becoming more resource efficient companies will gain a competitive advantage. Their costs will be lower enabling them to sell at a lower price. It will also help the company to be more resilient to any shortage or increase in the cost of raw materials.

It might appear tempting to increase taxes on resource extraction and energy used by companies. This would encourage investment in energy and resource efficiency. The problem is that if one country, such as the UK, increases such taxes whilst others do not, then British companies will be at a competitive disadvantage and may lose sales or move their operations overseas. In theory we could avoid this by establishing a web of tariffs on imports from countries who do not tax energy and resources. This would quickly get complicated and cause endless international trade disputes. So, governments should increase taxes on resources where possible, but they cannot push it too far unless other countries follow suit.

Government can support the innovation necessary to enable society to reduce its carbon emissions and its impact on wildlife. As a simple example, by increasing productivity of growing crops, we need less land to grow food and it should be possible to return some land to other uses such as for wildlife. Companies are now experimenting with growing food indoors, so called vertical agriculture, where the energy to grow plants comes from LED light bulbs. This can be extremely productive with up to six harvests in a year, no water wasted, and no fertiliser or pesticide required. This could enable year-round and local production of certain high value foods such as salads. Governments can encourage and influence such company innovation through grants and tax breaks. These can be generic or targeted at specific areas of identified need, such as more efficient batteries for electric cars and electricity storage. Alternatively, large companies can specify their innovation needs and invite universities and small innovative companies to collaborate with them to tackle and solve their challenges.

We need to do everything we can to accelerate and drive innovation towards the goals of reducing society's carbon emissions and impact on land-use and wildlife.

****

All companies can directly benefit from investing in energy and resource efficiency. Companies can also shift their business focus towards supporting a wider low carbon economy. The Scottish company, Weir Group, announced a strategic move from providing engineering solutions for oil and gas companies to targeting the mining sector – the metals such as lithium and copper needed for batteries and renewable generation. Alexander Dennis was a traditional builder of diesel buses based in Falkirk. They took a strategic decision to innovate and diversify into manufacturing hybrid diesel electric buses. Then they started building fully electric buses. Several years later, an increasing number of cities are banning diesel vehicles from their city centres to promote better air quality and this is leading to a surge in the demand for electric buses. Scotland has retained over 1,000 manufacturing jobs due to the far-sighted strategic decisions made by this company, supported by the local economic development agency. However, more recently Covid-19 has hit the growth of the business due to the drop in passenger numbers.

Multi-national businesses seem to be competing to announce ever more ambitious carbon cutting and sustainability targets. Many should be treated with a degree of scepticism. To be truly sustainable requires support from the directors, training and a change in culture across the organisation, backed up with money to invest in innovation and new technology. Governments should tax the use of virgin raw materials and regulate far harder to support the implementation of a circular economy. We need individual businesses and the trade bodies to stop dwelling on the past and campaigning to block or delay necessary change. Instead, they should collaborate with consumers and government to develop sensible regulations and to focus on maximising the new opportunities which arise from any change. Business can be a force for good, influencing consumers and its supply chain to cut emissions and to enhance the environment and still make money. A positive vision for the future would be factories that use recycled or re-useable raw materials, run on renewable energy, create no waste and sell long-lasting quality products that can be repaired and reused. Is this not better than what most businesses and retailers serve us with now?

Chapter 25:
Rewilding Nature

Wildlife and nature are fundamental to human livelihoods and wellbeing. I also believe it is morally right to preserve and provide habitat to allow wildlife to thrive and that active or passive contribution to the extinction of a species is morally wrong. 'Nurture nature' is clearly the most relevant of my five common-sense principles to this chapter. It is not about how we protect and conserve the remaining wildlife; it is a positive message to restore and enhance nature – and to improve our lives.

Traditionally we have 'conserved' wildlife and nature, protecting the best bits that remain in designated nature reserves. Reserves are actively managed to provide habitat for selected species. Sometimes this means keeping wildlife and people apart or restricting access to certain areas. Rewilding turns this on its head. Natural processes dominate allowing nature and species distribution to evolve over time. Rewilding applies to rural areas and to our towns and cities and can envisage people living with nature. It is a positive vision for the future.

Our economic system hugely undervalues nature and wildlife. A landowner can cut down trees and sell them for profit. There is no cost placed on the loss of wildlife habitat or economic losses. The exposed land is more prone to erosion causing rivers to silt and deforestation can lead to flooding of properties downstream - but these costs fall on someone else. Similarly, a farmer can use metaldehyde-based slug pellets which are slightly cheaper than alternatives, but they pollute watercourses costing the water utilities millions of pounds to attempt to remove. If we have the will to do so, we can tackle such pollution, including emissions of carbon dioxide, through regulations or taxes. It is much harder, and sometimes impossible, to put an economic price on nature – raw materials, soil and flood protection, insect pollination of crops, a source of new pharmaceuticals, joy and aesthetics. Many have tried to value these 'ecosystem services' but even if you calculate a price it is difficult to integrate it into our economic decision-making system in any meaningful way. The UK's Aggregates Tax is one small example of a tax on destructive extraction of raw materials from nature, but it is difficult to apply sensible taxes across all our impacts on nature. The price mechanism is never going to satisfactorily protect nature, so we need other tools to do so.

The most obvious way to protect nature is through regulation, for example, creating area-based designations such as national parks, nature reserves and Sites of Special Scientific Interest. Within these areas, regulations protect existing wildlife, and hopefully enhance it. But area-based designations are not enough as this creates 'islands' of protected wildlife often surrounded by a 'sea' of intensive agriculture. Over the long-term, this strategy is doomed to fail, as physical isolation of creatures will inevitably lead to genetic stagnation and increased susceptibility to disease. So, we also need wildlife corridors to link nature reserves. Moreover, to enable wildlife to thrive, we need policies and regulations that apply across all the countryside. The widespread use of pesticides has led to significant a decline in insects on farmland, but also adversely impacts neighbouring nature reserves. Excessive use of fertiliser can lead to run-off of nutrients into rivers and damage wildlife downstream. Everything is inter-connected.

For wildlife to thrive it needs space and lots of it. In his book Half Earth, Edward O. Wilson, proposes setting aside half of our planet for wildlife. Humans can just about survive with one kidney and one lung, but we would not thrive with half of our vital organs. In the same way he suggests that half of

Earth is the minimum that nature needs. Even at this level, many species will become extinct. At present, around 15% of land and 7% of our oceans benefit from some form of environmental protection status. In the USA 14% of land has some level of environmental protection and 12% of its seas. Venezuela, Slovenia and tiny Monaco have already designated more than 50% of their land although this does not always mean that such protection is effective or that there is no exploitation across these areas. Costa Rica, one of the most successful 'green' economies in the world has protected 28% of its land.

The scientific basis for protecting large areas is to address the fact that habitat loss is the primary driver of wildlife loss. Human activity has caused the extinction of at least 680 vertebrate species since the year 1500, many on isolated islands. Many more, small, creatures without a backbone have also vanished but have not always been recorded. More worryingly, the projections are that one million species are at risk from human activity including destruction and conversion of habitat, hunting, pesticides, the spread of invasive species and climate change.

I support the principle of half Earth, albeit the reality is complicated. We will need a healthy mix of areas for wildlife, areas for intensive agriculture and areas where human activity and wildlife can thrive together. We will need a shift from meat-based diets and for agricultural productivity to grow faster than human population growth. We should not just set aside land that is unproductive for humans – the ice caps, deserts and land of poor agricultural value. For half Earth to work properly it needs to include a mix of all habitats and climatic zones including the biodiverse tropical rain forests and some productive agricultural land in temperate zones.

****

Forests, peatlands, mangroves and permafrost store carbon. Peatlands cover 3% of the Earth's land surface but store one-quarter of land-based carbon.   By destroying these ecosystems, we release more carbon.   Conversely restoring them will help to remove carbon from the atmosphere. This is the principle behind some carbon offset schemes, where you pay an agreed amount per tonne of carbon emitted, and this will be 'balanced' by restoring peatland or forest elsewhere. I am sceptical about some aspects of carbon offsets - more on this in the next chapter - but it could certainly bring in substantial funds to restore nature and provide better habitat for wildlife.

In the UK we have destroyed 98% of our natural woodland. Drainage and land-use changes have destroyed most of our wetlands and degraded much of our peatland. We still allow householders to burn peat in open fires and the horticultural trade extracts it as a potting agent for young plants. Carrying this historical baggage, it is difficult to lecture countries like Brazil and Indonesia who are copying what we did and rapidly replacing forest with agriculture to help them grow their economies and raise living standards. The honest approach is to admit to our mistakes, to repair some of the damage we have caused, and from there to promote sustainable economic and social models that will also protect forests and wildlife.

The national parks are at the core of the UK's conservation efforts, but we farm or manage most of this land.   For example, much of the Cairngorms, the remotest and wildest national park in the UK, is managed for grouse shooting. In most upland areas grazing by sheep or deer prevent natural tree regrowth. In other countries it might be goats. Unlike some around the world, British national parks were all established in areas that humans had lived in for centuries. The government requires UK national park authorities to conserve nature; and to promote the sustainable use of resources, education, recreation and sustainable economic development. These are all worthy aims, but sometimes compete. A balance between human activity and wildlife is important, but the overarching ambition in these protected areas should be to improve the habitat for wildlife.

We need a massive shift in the balance of our land-use. Large scale rewilding can restore naturally functioning ecosystems which in many cases would be more productive than the current land-use. We should license shooting estates and impose a legal responsibility to promote natural regeneration. Just like companies should have a social license to do good for society, landowners privileged enough to own large estates should be custodians of that land and have a statutory duty to improve the land for nature.

Small steps have already been taken to rewild certain areas in Scotland, trying to recreate natural processes before modern humans altered or destroyed it. Intrusive timber plantations in the Caithness Flow Country have been removed to restore the natural peatland. In Glen Affric and in the Trossachs, areas are fenced to keep grazing animals out. Native trees have been planted and after only a few years birdlife returns. In fact, once the grazing pressure is reduced, trees start growing spontaneously after three or four years. Artificial obstacles to fish migration in rivers such as weirs are being removed. Beavers have been reintroduced to forests in Knapdale and the River Tay catchment. Like elephants and ants in Africa, they are a 'keystone' species, which manipulate and alter the landscape. They gnaw through small deciduous trees to build dams. This creates temporary clearings and wetland which create new habitat for insects, fish, amphibians and birds. Wetlands store more water which can reduce peak-flows and flooding downstream, although farmers do not always welcome these new wetlands on their land.

We should restore much of this natural forest, focusing on areas that are of marginal use for agriculture, but also some areas of prime productive land. And it is not just forests; this restoration applies to rivers and wetlands, wildflower meadows and our seas and estuaries.

Imagine if we restored a vast Caledonian woodland habitat in the Highlands of Scotland stretching as far as the eye can see. There would be free flowing wild rivers and wetlands, moorland and mountains rising above the natural treeline. Fish, birds, amphibians and mammals could thrive again. Large grazing animals and predators could be reintroduced. Certainly, the lynx, perhaps bears and wolves but only once the restored area is large enough. Wolves are making a comeback in mainland Europe, including densely populated Belgium, so why should Scotland be any different, other than we have destroyed the natural habitat that they can thrive in? Wolves were reintroduced to Yellowstone National Park in the USA in 1995. The presence of this predator makes deer nervous and constantly on the move. This reduces grazing pressure and enables trees to flourish. Woodland bird numbers increased as did beavers that flourished on the new tree saplings and created new wetland habitat. Mice numbers increased and eagles that prey on them. Bears ate the berries on the new shrubs. Even the rivers changed character as new trees helped to stabilise the banks and reduce soil erosion and sediment.

This Scottish Caledonian woodland habitat would not be a complete wilderness, but there would be space for wildlife to move around within it, with woodland corridors linking it to other restored forests across the UK. And it would not be continuous forest. Herbivores, large and small, have different grazing habits that create a mosaic of open and forest habitats, whilst their manure spreads nutrients and seeds across the countryside to benefit plants and insects.

Rewilding is not just about the land. Many rivers and seas have been devastated by pollution and over exploitation.

Whale hunting is an obvious example but there are many others. Off the Scottish coast, dredgers have scoured the seabed for scallops, effectively raking the seabed and disrupting everything that lives or grows there. This is short-termism by fishermen as it then takes much longer for the scallops to recover. The Firth of Forth had the largest oyster beds in the world in the 18th century. The development of heavy rakes dragged along the seabed enabled people in rowing boats to collect 30 million oysters in a single year. This overfishing, and the lack of any effective management, led to a collapse in the oyster population and eventual extinction from the east coast of Scotland. More recently, the government has established small Marine Conservation Areas. Following a community led campaign; a small 'no-take' area was established off the island of Arran in 2008. Studies have shown that wildlife is recovering naturally. King scallops and lobster density is higher, carbon absorbing seagrass has returned to the seabed and the area is now a nursery for juvenile fish, especially cod. By protecting even a small area the benefits accrue over a wider area which can enable sustainable commercial fishing to take place.

The Forth Rivers Trust has taken small steps to transform the Allan Water, a river in Scotland. The riverside landscape is picturesque, but ecologists and hydrologists will notice that after rain, its waters turn muddy as soil and peat washes off fields and moors. This silt can harm fish such as salmon. Humans have completely altered the river's catchment. Farmers and landowners have removed the natural forest cover, drained peat bogs and canalised stretches, whilst invasive species such as giant hogweed are rampant. Now the Trust is blocking drains to raise the water table in peatland high up in the catchment. This enables moss to regrow – in effect acting as a giant sponge. Invertebrates, amphibians and birds are returning to these bogs. Trees planted along the riverbank shade fish, stabilise the banks and reduce soil erosion. These actions help to store carbon on the land and reduce the risk of flash floods. Attempts are being made to control invasive mink and giant hogweed. There are plans to remove barriers to fish migration and to restore the original meanders, in effect to rewild the catchment. But lack of funds limits action on the ground.

In large parts of rural Europe, the USA and China, young people are abandoning remote, hilly and marginal land as they migrate towards cities. This provides an opportunity to rewild areas to benefit wildlife. Tanzania has suffered from years of deforestation, yet some land has reverted from agriculture back to forest which is helping to recreate the habitat that chimpanzees need. Other more deliberate efforts are being made to restore forests. Satellite images that reveal the full scale of forest loss in Tanzania are used to collaborate with local communities to encourage them to establish forest reserves and set aside some woodland for gathering firewood. Communities benefit from soil stability, fewer landslips, cleaner rivers and reduced floods whilst wildlife tourism provides a new source of income.

Intensive agriculture is the prime principal driver of wildlife loss. Monoculture replaces natural habitats, with land drained, streams straightened and trees and hedgerows removed. Insecticides eliminate harmful pests from crops, but clearly, they also kill other beneficial insects. Herbicides can be targeted at certain weeds, but this means they cannot flower and provide pollen for insects. Birds and bats rely on these insects. Wildlife has been pushed into 'islands' or 'oases of disconnected nature reserves and lack many natural processes such as predator and prey relationships. In Europe, subsidies designed to maximise production have driven the intensification of agriculture. For example, a subsidy based on the number of sheep kept will naturally encourage farmers to overgraze a landscape. We should redirect these subsidies to support wildlife - replant trees and hedgerows, not allowing animals to graze and defecate on riverbanks, restore rivers to flow more naturally, create wildflower meadows and discourage fertiliser application at certain times of year when much of it is likely to run-off into rivers.

Meanwhile, invasive species continue their relentless march to conquer the planet. Wherever you live, if you investigate, you will find that invasive species reduce diversity and create havoc on wildlife and ecosystems. Landowners deliberately introduced South American cane toads to control insect pests, but they are spreading uncontrollably across Australia. Their toxic glands are highly poisonous to native wildlife that would otherwise control their spread. In Florida, Burmese pythons are spreading across the Everglades reducing the population of native wildlife. Zebra mussels, from south Russia, have invaded the Great Lakes. A female can produce up to one million eggs enabling them to spread rapidly. In addition to adverse impacts on native wildlife, the mussel shells can clog up pipework and water treatment plants. Lionfish, from south-east Asia have spread across the Caribbean. They are prolific breeders, have a voracious appetite, hoover up all small fish in their path and outcompete native predators such as sharks. Of even greater concern are small species, for example New Zealand flatworms invading parts of Europe. They eat native worms and are less beneficial than native worms at helping to make soil productive. The spread of small species, including most invasive insects, is physically unstoppable with current technology.

Wildlife on islands can be particularly susceptible to damage by non-native species such as rats and even hedgehogs that eat bird eggs. However, there is an opportunity on islands, with concerted effort to eliminate such invasive pests. A well-funded campaign has successfully eliminated rats from the island of South Georgia. In the Western Isles of Scotland, hedgehogs were initially culled, later relocated to the Scottish mainland away from bird nesting sites. Birds in New Zealand evolved without any mammal predators and were decimated after humans introduced rats, dogs, cats, stoats, weasels and possums. Now there is an ambitious and long-term campaign to rid the entire country of rats, possums and stoats by 2050.

Regulations to stop the international spread of non-native species including pests and diseases are woefully inadequate. Only a few countries, such as Australia, take it seriously, learning from their early failed battle against the destructive spread of rabbits. In the Mediterranean there are no regulations to control the spread of marine invasive species. Organisms such as algae, molluscs, crabs, barnacles and jellyfish can attach to the undersides of boats whilst ballast water enables organisms to hitchhike a lift across oceans. Leisure marinas are a growing problem, especially yachts that travel long-distances by sea or on trailers. The only effective solution is tighter regulation, a risk-based monitoring and inspection system and regular cleaning of the undersides of boats. Then, we need to deploy a rapid reactive force to tackle any reports of new non-native species before they spread and become impossible to control. All this will cost money and will just slow their spread.

Controlling invasive species is time consuming and expensive. It requires good organisation and perseverance. In Scotland grants have been awarded to eradicate giant hogweed, but as soon as the money runs out, the spraying ends and the plants quickly re-establish. The Scottish Government's wildlife body, Scottish Natural Heritage develops strategies and monitors the spread of non-native species, but the main responsibility to control them is down to a myriad of landowners and cash strapped local councils who do not treat it as a priority and consequently do not make the necessary funds available.

The Scottish Wildlife Trust is leading a partnership to stop and reverse the spread of grey squirrels. These invasive animals from America out-compete the native red squirrel. They are immune to squirrel pox disease, but they can transfer it to infect and kill red squirrels. And grey squirrels strip bark preventing trees such as oaks from thriving. Volunteers now trap grey squirrels to enable red squirrels to re-colonise and flourish again. The recent spread of the pine martin may also help as they find it easier to catch grey squirrels in tree branches than the lighter and more nimble red squirrels.

This is an area that would benefit from community led action. Communities can take a long-term approach and follow up to ensure invasive species do not re-establish. The UK's Environmental Audit Committee called for an 'army' of one million volunteers to monitor and tackle invasive species. It is not clear who is going to organise this, but the volunteers will need some funds, training and equipment.

Educating people is at the core of the rewilding concept. Most of us are divorced from nature and ignorant of where our food comes from. We do not understand the complex web of life, and how impacting one area or species may have adverse knock-on effects. We might enjoy nature documentaries but are we aware that our excessive consumption puts land under pressure? We should start with our children. At primary school children should learn about nature. Much of this could be hands-on in the countryside, ranging from finding and identifying species to practical conservation work to improve habitats for wildlife. Children should visit farms and meet farmers and landowners. Secondary schools should teach all pupils ecology, with the option of advanced exams for some. This would encompass biodiversity, plant succession and the interaction between species; habitat loss, non-native species and other threats to wildlife; and visits to inspiring examples of rewilding projects that engage with local communities.

Children should learn about the value and uses of nature. A single species of tree, such as the silver birch, is much more than an attractive addition to the landscape. Birch trees support a large community of insects and other invertebrates, with 330 species known to feed on them. These invertebrates are food for birds, and siskin feed directly on the seeds. Red deer browse on the leaves of young trees. The roots have symbiotic relationships with fungi, including chanterelle which foragers can collect to eat. Primroses, violets and bluebells flower in early spring, before new leaves limit the light reaching the forest floor, but enough remains to enable edible blaeberry and cowberry to flourish beneath them in the summer.

Birch is a hardwood that can be used directly in construction or to make wooden crafts. Its fine grain makes high quality furniture. It is flammable, and its high calorific value is good for firewood. It can be made into plywood for construction, and its flexibility is a good property for skateboards. Its bark is waterproof and can be used to make traditional boats and canoes. It contains betulin that the pharmaceutical industry can use, and other chemicals to make perfume and soap. Birch can be used to make paper although it is not as high quality as other trees. Its leaves effectively absorb air pollution along busy urban roads. Its trunk can be tapped to collect liquid sap full of antioxidants and vitamins, a 'superfood' in some people's eye. We can drink this or use it as a healthy alternative to refined sugar or eat with pancakes instead of maple syrup. The sap can ferment to make wine or beer and its leaves can flavour tea. The chaga mushroom is a parasitic fungus that grows on birch trees. It is full of antioxidants and can be processed or fermented into a tincture - a concentrated herbal extract then drunk as a tea or fruit juice with purported health benefits. All of this from one species of tree – and there are 60,000 tree species in the world.

A restored landscape and wildlife would be good for people too. Being close to nature is the antithesis of an indoor, sedentary, screen-based lifestyle. Children benefit and develop from being close to nature, and access to nature is good for the mental health of adults, particularly important in deprived city areas. Wild landscapes already draw tourists to locations such as the Scottish Highlands, but these are empty landscapes – empty of people and empty of wildlife diversity. A restored landscape would teem with wildlife including reintroduced beavers, sea eagles and pine martins. A new tourism industry could be built around wildlife, rather than just scenic landscapes. In Scotland, visitors already come to visit the fish ladder at Pitlochry to view salmon, puffins on the islands, ospreys at Loch Garten, red kites at Argaty near Doune, dolphins in the Moray Firth. The nature of wildlife tourism could change from visits to isolated strongholds of wildlife to nature safaris on land and sea. A wild and natural landscape would attract other tourists including walkers, cyclists, photographers, canoeists and anglers.

Commercial forestry, deer stalking, pheasant shooting, and fishing could continue. Limited numbers of cattle and sheep could provide more diverse habitat for wildlife. Woodcraft skills, craft workers, joinery and biomass for burning could all expand. The more attractive landscape might attract people who can work from home such as those with information technology skills.

Many tourism and food and drink companies already benefit from Scotland's image as a clean, green location. Hotels and restaurants can boast that their menus offer customers fresh, local produce. Rewilding would enhance Scotland's reputation enormously, offering salmon, shellfish, venison, wild berries, herbs and mushrooms. Whisky thrives on Scotland's reputation of a clean environment and clean water. Many distilleries are open as tourist attractions and Glenmorangie have built an anaerobic digester to treat the effluent from manufacturing. They are rewilding the sea by working with Heriot Watt University and the Marine Conservation Society to restore oysters to the Dornoch Firth. Oysters filter and clean the water and create beds that provide a habitat for juvenile fish and invertebrates.

Humans need nature and given the state of human influence on the planet, nature needs humans to protect it and to give space for wildlife to thrive. We continue to undervalue nature and wildlife; in a geological timeframe destroying much of it in the blink of an eye. We should learn to work with nature and work to change attitudes. Trophy hunting of big cats was a source of pride not so long ago, now frowned upon. The Chinese authorities now discourage the use of rhinoceroses' horns for traditional medicine. More comprehensive education would help people to understand the value of nature. Rewilding is slowly gaining an unstoppable momentum, restoring nature but also working with it to maintain livelihoods. As crop agriculture becomes more productive, and if we shift from a meat-based diet, this could open large areas of land to rewild to benefit humans and nature. Companies paying into carbon offset schemes could pay for much of this. Meanwhile, governments should redirect subsidies to farmers and landowners to enhance nature.

In densely populated countries this rewilding is unlikely to be a simple matter of abandoning vast tracts of land to nature. That might result in an impenetrable, less diverse landscape. Humans need to value and work with nature. Our cities can be greener, with walls and roofs covered in vegetation, but also bluer, with streams and ponds providing attractive features. There would be large protected nature reserves; vast areas that are wild, many more that would be managed for humans and nature with wildlife corridors linking them. Farmers would grow food but also work to improve the soil, limit pesticide and insecticide use and allow wildlife to thrive, providing a better environment for us all.

Chapter 26:
Carbon Offsets

What if we could remove carbon from the atmosphere and store it safely and permanently? This could compensate for the carbon that it is difficult or very expensive to stop emitting. In future, in theory, we could use it to reduce the volume of carbon dioxide in the atmosphere. Of my five common-sense principles, 'price carbon pollution' is most relevant to this chapter, followed by 'be fair across current and future generations' and 'nurture nature'.

There are several ways that humans can remove carbon, but firstly it is useful to consider the total emissions that an average person living in the UK might be responsible for over their lifetime. This will depend on when they were born, with a lower figure for young people - increasing affluence has tended to increase emissions over time, but in the UK at least, this is more than offset by the long-term decarbonisation of the economy.

My rough estimate for a 50-year old person in the UK today, expected to live to 80 years is that they will create 1,000 tonnes of carbon dioxide equivalent over their lifetime (20, 15, 10 and 5 tonnes/year for 20 years each). Affluent people will emit much more. Assuming we reach zero carbon by 2050, a 20-year old today, also living to 80, might be responsible for less than 500 tonnes (12.5, 7.5, 2.5 and 0 tonnes/year).

Is it possible and credible to cancel out all this harm by paying into a carbon offset scheme? It would be far better to cut our emissions rather than to pay to offset them but, firstly let us consider the cost. Many schemes currently offer carbon offsets at less than £10 per tonne. Emitting one tonne of carbon under the European Emissions Trading system currently costs 25 euros, around £20. Given complexities, including the falling price of renewable energy, economists cannot agree on the price needed to properly tackle our emissions. But most suggest that it should escalate over time, perhaps to over £100 per tonne. So, it might cost between £5,000 and £100,000 to cancel out your lifetime emissions. My gut feel is that it needs to be towards the upper end of this range. This sounds a lot, but is it a high price to protect our planet, your grand-children and wildlife? It equates to between £1.20 and £24 per week. To put this into context, smoking a packet of 20 cigarettes per day (£10) costs £3,650 per year or £219,000 over 60 years, and this is not even the level of a heavy smoker. The alternative might be to live on a planet devastated by climate change.

****

Carbon sequestration is the removal of carbon dioxide from the atmosphere. There is a short-term biological carbon cycle in the atmosphere, oceans, soil, vegetation and freshwater dominated by photosynthesis and respiration of plants and phytoplankton. Then there is a long-term carbon cycle over geological timescales dominated by chemical weathering of rocks, sedimentation and volcanoes. Some processes turn the short-term into the long-term. For example, plankton near the ocean surface, die and sink burying their carbon into deep ocean sediments, whilst coal is dead vegetation that did not rot, and became deeply buried and compressed. This complexity and variety provide us with a range of options to enhance and encourage these natural processes. Other options include capturing carbon from power plants to use in products, bury carbon underground in depleted oil fields, gas fields or other suitable rocks, or even to capture carbon direct from the air.

Trees are nature's ultimate carbon capture machine. They remove carbon dioxide from the air and store it within their trunks, branches, leaves and roots. Collectively trees affect the level of carbon dioxide in the atmosphere each year. In May, carbon dioxide peaks before the predominance of deciduous trees in the northern hemisphere soak some of it up in their leaves leading to a measurable dip around September. Trees store carbon dioxide until they die, then release it slowly back to the soil and atmosphere as they decay. Some trees can live for centuries whilst others are harvested after 20 to 30 years. If we burn the harvested wood, then this releases this carbon immediately. However, if we incorporate it into a building or furniture then it is locked away for further decades. In Sweden, researchers have built a 30-metre high wind turbine out of laminated wood, with plans for commercial structures 100 metres high. Not only do such turbines generate renewable energy, manufacturing them locks away carbon for decades.

****

We can make a useful distinction between a carbon store and drawing carbon out of the atmosphere. Protecting an existing forest simply stores carbon in balance; whilst reforesting an area of land draws down carbon. In effect, reforesting can restore us to where we were before deforestation occurred, but it cannot tackle the huge excess emissions from burning fossil fuels. The priority is to prevent further deforestation through conservation designation, regulations and carbon offset projects. Like many other countries, the UK is substantially deforested and imports 80% of its timber needs. There is ample scope for reforestation, much of it on land currently used for grazing. Commercial trees, like Sitka spruce, grow fast and absorb carbon quickly but the long-term benefit will depend on what happens to the wood after it is harvested. Conversely, an oak tree might absorb carbon slowly at first but will continue to absorb it for decades, even centuries. Forests will eventually reach a steady state, storing carbon, but not drawing down any more.

Lloyds Bank is investing in carbon offsetting to cut the carbon emissions from the activities it finances by half by 2030. They have entered into a £36 million partnership with the Woodland Trust, to plant ten million trees by 2030 which should absorb 2.5 million tonnes of carbon over a 100-year period. However, planting trees to capture carbon does have its complications, particularly in more remote regions that are difficult to monitor or are prone to corruption. Forest certification schemes help, but there are still questions around whether the trees would have been planted anyway, will neighbouring forests be cut down instead, will a monoculture be created, will the trees be harvested early? Then there is a risk that disease or a forest fire will wipe out any gains achieved.

Although we often think about the amount of carbon dioxide in forests and in the atmosphere, soils contain even more carbon. Poor management and farming practices can quickly degrade soil to hold less organic and carbon rich matter. Ploughing up natural grassland releases much of the carbon in the soil. Conversely, restoring degraded soils to better health can improve productivity and store more carbon. Soil carbon projects are not as popular as more visible activities such as planting trees, but the USA is trialling them. Their success is hard to measure, will vary depending on the soil and climate and are currently unproven over long timescales. Another option is to produce biochar – wood baked in a low oxygen environment to produce a carbon-rich material like charcoal. Farmers can spread biochar on fields to enrich acidic soils and absorb and retain moisture.

Peat is formed by the accumulation of partially decayed vegetation and organic matter. It forms in damp climates where the wet soil restricts oxygen, slowing the rate of decomposition. Peat formation slowly traps carbon in the soil over decades, but the carbon can be released from damage to peatland by drainage or extraction. It is not sensible to plant trees on peatland as the trees would soak up less carbon than is released from damaging the peat. Unfortunately, people are still setting fire to and draining natural wetlands, particularly in south-east Asia, to convert carbon-rich peatland to agriculture or palm-oil plantations.

Many other natural habitats are also effective at storing carbon. Seagrass absorbs carbon even better than tropical forests and provides a good habitat for juvenile fish. Similarly, salt marshes, mangroves and coral reefs absorb carbon. They also create diverse habitats and protect coastlines from storms.

Soil and habitat restoration often have multiple environmental benefits and are relatively cheap methods of storing carbon, but most of these stores will become saturated after a few decades and there is simply insufficient land available to offset all the carbon necessary.

An alternative is to enhance natural processes that remove carbon dioxide from the air. Reykjavik Energy in Iceland uses reactive materials to bond, or mineralise, carbon dioxide, then inject it to form carbonates in the pores of underground rocks. An alternative is to spread pulverised basalt rock onto agricultural land. Weathering produces an alkaline bicarbonate that improves soil fertility and will later be washed into the sea for permanent storage. Oceans could be fertilised by adding iron to encourage phytoplankton to grow, trapping carbon dioxide in ocean sediments when they die. These are all possible, but all have side effects. They require extensive mining, energy use and may have unintended adverse impacts on ecosystems. We need further research and trials, but none will be rolled out at scale without an effective price on carbon pollution that creates a market for business to deliver.

Carbon capture from power stations or major industrial users is an emerging, but expensive technology promoted by the coal and gas industry. The oil industry has used it for decades. Cleverly, or ironically depending on your viewpoint, they pump the captured carbon dioxide back into oil fields to increase the pressure to force out a higher proportion of the remaining oil contained in pores within rocks. For electricity power stations fuelled by coal and gas, carbon capture requires chemicals and substantial additional energy to operate. Further energy is used to pump the captured carbon dioxide to a suitable underground storage area. It is this inherent inefficiency, with no additional benefits, which makes this an unattractive process. However, gas power stations may be necessary to help balance regional electricity grids so there may be a limited role for carbon capture and storage from these plants. Carbon capture may also be a useful technology to decarbonise industrial processes that require high temperatures such as manufacturing steel and cement. Furthermore, carbon capture and storage from a power station fuelled by biomass could be carbon negative, removing and storing carbon deep underground. Given the need to offset greenhouse gases from aviation and agriculture even in a low carbon economy, this option is proving attractive to policy makers in countries aiming to be net zero. Again, this does have a place, but is subject to limits on the supply of suitable biomass.

Aberdeen based company, Carbon Capture Machine, is experimenting with using industrial waste carbon dioxide to produce precipitated calcium carbonate and magnesium carbonate. These are raw materials commonly used to make paper, plastics, paints, adhesives and even toothpaste. They could even be added to cement and concrete with the carbon locked away for decades. Many other companies are working to produce useful products, or even liquid fuels, from carbon dioxide, but none are commercially viable with the current low price and limited market for carbon. It is also unclear whether they can be sensibly scaled up given most require mined chemicals and use energy in the process.

It might also be possible to remove carbon dioxide directly from the atmosphere, a process called direct air capture. In theory, this could be situated anywhere; perhaps beside a source of renewable power or near a suitable storage location. However, carbon dioxide emitted from a coal power station may be one molecule in ten, whilst it is only around one in 2,500 in the wider atmosphere, requiring far more energy to capture. This process requires minerals to capture the carbon and would have to use a low carbon energy source otherwise the net benefits would be minimal. It would often be better to use this low carbon energy source for other purposes such as to make hydrogen. Although companies continue to develop direct air capture technologies it seems unlikely that this will be a sensible solution to climate change in the near future. In any case, nothing will happen at scale without people willing to pay a very high price to remove carbon from the atmosphere. In the meantime, it is much better to avoid emitting carbon to the atmosphere in the first place.

****

Many individuals, companies or governments want to offset their residual carbon emissions - those that are difficult, expensive or impossible to avoid. They pay an agreed amount per tonne of carbon and in return the offsetting organisation promises to remove (or avoid) an equivalent tonnage of carbon somewhere else. Typically, this is through investing in energy efficiency, building additional renewable energy capacity, planting trees or restoring peatland. It can apply to methane too, such as capturing leaks from landfill sites or abandoned oil wells. Each of these investments has slightly different impacts. Energy efficiency should reduce carbon emissions, renewable energy is likely to prevent an increase in emissions, planting trees should remove carbon for several decades, restoring peatland should store it for millennia, and reducing methane leaks avoids a strong greenhouse gas from entering the atmosphere.

Carbon offsetting often takes place far away, out of sight of the company or individual who is paying for it. An alternative for companies is carbon 'insetting' where they invest in energy efficiency projects within their own company or in their supply chain. This can be more accountable, enhance their reputation, and reduce their costs or directly assist their suppliers. For this to be successful and credible, it requires collaboration and transparency. Meanwhile, companies or governments who are fortunate enough to own land can invest in carbon reducing projects on their land. An example is Scottish Water, a utility that owns vast tracts of peatland which they could restore to offset carbon and to enhance water quality within their catchments.

Carbon offsetting has gained a poor reputation amongst many environmental campaigners. The main argument is that it can be an easy option by a company to claim carbon neutrality and avoid the tough decisions and actions needed to actually reduce their emissions. A company or industry body can also use it as ammunition to counteract regulatory pressure to cut emissions. The international aviation industry's proposals to offset the growth in emissions after 2020 could be a tactic to stave off tougher regulations. For individuals, offsetting our emissions is potentially an easy and cheap method to ease our guilty conscience. We might use it to justify and enable us to continue our high carbon emitting lifestyles; driving long distances, buying a sports utility vehicle or flying long-haul. An alternative is for the government to place a sufficiently high tax on high carbon emitting activities such as long-haul flights. The money raised could be spent elsewhere to bury carbon dioxide permanently underground.

There are concerns about the environmental and wider sustainable development benefits of offset schemes. For example, planting a monoculture forest might reduce carbon but could have adverse impacts on wildlife. Conversely, some schemes such as supplying energy efficient cooking stoves to replace kerosene stoves have a carbon benefit and improve health through less indoor air pollution. But carbon offset schemes may not achieve their carbon saving objectives over the long run. For example, a major fire could wipe out the benefits from tree planting or peat restoration. In contrast, carbon capture and storage does have the advantage of permanently locking away carbon deep underground. To be effective, carbon offset schemes should also be additional to what would have happened anyway. If a government regulates to require all new light bulbs to be energy efficient, then providing energy efficient light bulbs under a carbon offset scheme simply supports an existing government policy and is not making any additional cuts to emissions.

The final argument against carbon offsetting is around the price paid to offset a tonne of carbon. The cost charged by companies offering to offset emissions varies widely, but is typically very low, around £8 per tonne of carbon. At this rate, assuming no adjustment for the additional climate impacts of aviation discussed earlier, it would add just £20 to the cost of a return flight from London to California which is too low to affect our behaviour. This price is based on the 'low hanging fruit' – projects that companies can currently deliver cheaply to reduce carbon emissions. After all, energy saving light bulbs are cost effective without subsidy; and in some places, it is cheap to buy land to plant trees. This does not mean that we can miraculously solve climate change through carbon offsets at £8 per tonne. If demand rose, then the low-cost projects would quickly be taken, and we would then have to invest in more expensive carbon saving projects. It is therefore disingenuous to suggest that we can truly offset our emissions at a low price as this only applies for a limited time and if a limited number of people choose to do so.

The problem is that the carbon offset market is unregulated. There are elements of a 'race to the bottom' where companies can make dubious claims about offsetting emissions at a low price. From a consumer's perspective of course, a low price seems attractive, but a low price will never be enough to pay for the more difficult and expensive projects that we need to prevent climate change.

If you are wealthy enough then there are other alternatives to paying into a carbon offset scheme. You could invest to reduce the carbon emissions that you directly control, for example insulate your home, switch from petrol to an electric car or install solar panels on your roof. If you have spare cash available, and can accept taking some risk, you can invest in green bonds or into an ethical fund that invests on your behalf on projects which cut carbon. Many investments in renewable energy have long payback periods which may suit long-term saving for a pension. It is even possible to invest directly into company projects, or renewable energy schemes, often through a crowd funding intermediary. This way you can see how your money is being used to reduce carbon emissions. If it is a local community run investment, such as Edinburgh's 'Solar for Schools' project then you can see your investment in action. Of course, such direct investments are risky, so never invest too much of your spare cash in them.

****

Reducing your emissions is always better than offsetting. Carbon offsetting is complicated, but on balance, it is better to use carbon offsets than the alternative of doing nothing. However, the best option is to reduce your emissions and still fund environmental projects for altruistic reasons.

There is an opportunity to combine the desire by some businesses and individuals to offset their emissions with the landscape scale conservation and rewilding that we need to restore wildlife. We should use the money raised from offsets to subsidise farmers and landowners to restore landscapes. Specifically, to plant new areas of woodland, to restore peat bogs and coastal marshes, and to change farming practices to increase the amount of organic material in soils.

For carbon offsets to be effective it is important that the market is regulated to ensure high standards. An internationally recognised accreditation standard would provide assurance to investors. Most importantly, a sufficiently high price should be levied. Ideally this would be set by an international body such as the United Nations, but if this is not practical, then the UK Committee on Climate Change could set an appropriate price. If this regulated price is higher than the 'market' rate, then more than one tonne of carbon could be offset when paying for one tonne. Once a realistic price is established, a nationwide carbon offset scheme should be set-up. There should be a compulsory scheme to offset certain high carbon emitting activities such as long-haul aviation and a voluntary scheme offering individuals the option to 'cancel all their carbon'. There would be the option to offset your current year's carbon, or for those who can afford it to offset their lifetime emissions. Who knows, this might attract some people when writing their will. Finally, climate change will most affect those in low income countries, but people in high income countries have caused most emissions. There is therefore a strong equity argument to spend some of the carbon offset money in low income countries if this is monitored and not at risk of corruption.

Chapter 27:
Covid-19 Consequences

This chapter was not on my radar as I began to draft this book early in 2020. Our understanding of Covid-19 is developing rapidly and our responses to it will evolve over time. I fear that whatever I write will be out of date by the time you read this, but I will try to explain the impacts that Covid-19 could have on our carbon emissions and on our relationship with nature.

There are two distinct categories of pandemic caused by infectious disease - viral diseases such as Covid-19 and the 1918 Spanish Flu; and bacterial diseases such as cholera and the Black Death. Two-thirds of these pathogens originate in animals, for example it is thought that Covid-19 originated in bats, SARS from bats and civet cats, Ebola from African fruit bats, and H1N1 swine flu jumped from pigs to humans in North America in 2009.

Normally pathogens infect their hosts, but do not kill them, as death will decrease the rate of transmission to other animals. But in crowded conditions, such as intensive livestock agriculture, this constraint no longer matters as the pathogen will never run out of hosts, so it is possible for more deadly diseases to develop in dense populations. One strain of H5N1 avian influenza in China poultry farms killed 60% of humans it infected, Ebola kills 50%, whilst Nipah virus originating in Malaysian pig farms killed 40%. Luckily Covid-19 did not originate from livestock and has a low mortality rate especially amongst young people.

Covid-19 shares 96% of the genome of a virus common in bats, but scientists believe that the virus must have mutated via an intermediary animal as its biology would not allow it to jump straight from bats to humans. Nipah virus jumped from fruit bats to pigs to humans, possibly arising from new agricultural plantations forcing fruit bats to move from the forest to an area of human occupation. A pangolin, a South American anteater, has been suggested, but not proven, as an intermediary for Covid-19. It is not known how Covid-19 jumped to humans but the first cases were identified in a fish market in Wuhan. These 'wet' markets are common in south-east Asia. Their crowded open-air stalls sell fish, live chickens for slaughter on site and often wild animals for sale. Bats can be eaten direct or in soup. China has banned eating pangolin, but there is still an illegal trade to eat it as a delicacy and its scales have purported medicinal virtues. Viruses can more easily spread if animals are sick, stressed or kept in dirty cramped conditions, such as stacked cages common in markets. Blood, urine and faeces can help to transmit disease.

Population growth, the spread of subsistence agriculture, habitat loss, the increase in bush-meat and the trade of captive wild animals all increase the unnatural interaction between humans and a wider range of animals. Climate change will exacerbate this as it will alter the range of animal species, pushing animals into new interactions with humans. These factors all increase the risk of future pandemics originating from animals. Across all countries we need tighter laws on the legal and illegal international trade in wild animals, eating bush-meat, wet markets and intensively reared livestock. And, governments need to enforce these laws.

For those living in western countries, we also need to consider that our intensively reared livestock and chicken farms are also a potential source of disease spreading from animals to humans. Spanish Flu, the deadliest pandemic on record, originated from birds. Now, there are 24 billion chickens, most kept in crowded conditions. The intentional breeding of animals to be genetically similar, the unnatural concentration of numbers, the routine use of antibiotics and mixing species create such risks. In effect, animals need social distancing rules too.

There are some similarities between the risk and impact of Covid-19 and climate change. Both crises are the result of an invisible enemy, whose danger accelerates over time. If you wait to see the effects, then it is too late to act. These make it difficult for politicians and society to take pre-emptive action. Like the dangers of climate change, scientists have long predicted the risk of a new pandemic arising from animals. Covid-19 may be a dress rehearsal for worse pandemics in the future and its disruption is an indication of the impacts of the climate related shocks that will come. And in both cases, pre-emptive action will help, but science and societal change can eventually solve these issues if we educate and invest in the right areas.

However, there are clearly differences between the impacts of Covid-19 and climate change. One is the immediacy of danger that forced politicians to react to the threat from Covid-19. Another is that the health risk from Covid-19 is primarily for older people, whilst young people face the most disruption to their careers and lives from both Covid-19 and climate change. There is no vaccine for climate change, but like tackling Covid-19 good science and early intervention will help.

Covid-19 does show that we can act decisively in the face of imminent danger. Individual governments have reacted to the crises in their own self-interest, but are beginning to take the World Health Organisation's advice more seriously and will need to cooperate to develop, manufacture and distribute a vaccine efficiently. To tackle climate change will also require international cooperation to share best practice.

****

The lockdowns in many countries and the collapse in global aviation have led to a significant, but temporary reduction in carbon emissions. During the peak lockdown in April 2020, global emissions fell by 17%. In countries with tight lockdown rules, like the UK, total emissions fell temporarily by 30%; mostly from transport - aviation fell by 90% and passenger vehicles by 60%. This led to a reduction in local air pollution in cities, by up to 60%.

The International Energy Authority's previously estimated an increase in global energy consumption of 1.3% for 2020. As a result of Covid-19, they have revised this to a 6% fall; oil down 10%, coal 8% and gas 5%. Renewable energy is up 1% as it is cheaper to operate and has therefore been fully used at the expense of other fuels.

Very roughly, global emissions arise equally from electricity, power used by industry, agriculture and land-use, and transport. For 2020, emissions from transport will fall the most, whilst electricity and heat will fall a little as more people work from home. Manufacturing and construction fell sharply but are likely to pick up again. Covid-19 has had relatively little impact on emissions from agriculture and land-use changes, such as deforestation. Most lockdowns did not start until March, so 2020 will have two or three months of normality resulting in an estimated overall fall in emissions over the year of around 8%.

Scientists estimate that we need to cut our emissions by 7.5% cumulatively every year this decade to prevent serious climate change. The lockdown has had severe and detrimental impacts on our lives so cutting emissions by use of a double lockdown in 2021 and triple in 2022 is not feasible. Also cutting our carbon emissions does nothing to stop worsening climate change. All it does is slow down the rate of warming. In fact, the change from the lockdown is miniscule. Imagine climbing up towards an erupting volcano. In 2020, we have not taken a step back from danger; we have simply slightly slowed our pace of ascent. In 2020, levels of carbon dioxide in the atmosphere are 416 parts per million. The previous prediction of an increase in carbon dioxide in the atmosphere of 2.8 parts per million in 2020 may only be 'reduced' to an increase of 2.5 parts.

Lockdowns have shown that changes in personal travel are important but are not enough to prevent climate change. We also need to change our diets, our consumption of goods and tackle the structural issues of electricity, heat production, iron, steel and cement manufacturing. These require further changes in behaviour and more government regulations.

Because of the steep reduction in flights and passenger car journeys the reduction in air pollution was far more dramatic than the fall in carbon emissions. This had an immediate benefit for millions of people, particularly those with respiratory difficulties. The pleasant results were visible to all - clear skies, better views, more visible stars and less noise enabling us to wake up to the sound of bird song. The uncongested streets and roads made walking and cycling a pleasant experience. The output of solar panels increased because of the reduction in haze and less dust settling on the panels. This gives us a foretaste of what a future world could be; one with clean, quiet electric vehicles, with cities designed so that fewer journeys by vehicle are necessary.

****

The reaction to the Covid-19 emergency by governments of all political persuasions has been like putting countries on a war footing, involving massive intervention in the economy and society. This proves that governments can take decisive action to tackle climate change if there is a political will.

The reaction to Covid-19 is well documented, the question is to what extent these changes will reverse or become permanent, and whether there are opportunities from this crisis that will also help in our battle against climate change.

Our reaction to Covid-19 has accelerated pre-existing trends in society. Home working is becoming normal practice, on-line shopping will continue to grow and the shift to a cashless society will be faster than expected. Business travel to conferences and for client meetings may never recover with a permanent impact on international business aviation. Previous tentative steps by schools, universities and health professionals to provide on-line services are accelerating rapidly, questioning the long-standing belief that we should always educate children and young adults in large centralised face to face groups. When SARS struck in 2003 it helped to kick start e-commerce in China. Similarly, but now globally, our use of information technology has accelerated because of Covid-19 restrictions on travel. Grandparents, who previously avoided digital technology, were forced to adopt the internet as the only way to see their loved ones. Businesses have quickly adopted laptops for home working, webinars to replace conferences and on-line training courses. Virtual reality technology will soon enable new choices, such as guided tours of remote factories.

These trends may cause lasting decline in the numbers travelling by public transport, the demand for office space and the survival of High Street shops. It may change where people choose to live. Some may choose to commute in single occupancy private cars to avoid proximity to other people on public transport.

Mining and burning coal were already under increasing pressure due to climate change concern and pressure from disinvestment campaigns and institutional investors. Covid-19 has highlighted that renewable energy is a safer investment.

Sweden closed its last coal fired power station early due to a fall in electricity demand, and the public has stronger evidence to support their campaigns for clean air. It is likely that 2020 will be the start of a long, slow, decline in the world's use of coal.

Covid-19 will accelerate digitisation and the use of new technology. Sensors connected to a control room can identify faults, for example on water pipelines, avoiding any need to travel to undertake inspections. Self-service touch screens and robots might replace restaurant waiters and hotel receptionists.

Confidence in international supply chains has been shaken. Long-distance, lean supply chains are inherently fragile, and can be affected by an earthquake, weather related disaster or a pandemic. There was a scramble to source protective equipment for health workers, and such insecurities of supply may encourage more local manufacturing. 3D printers can now manufacture protective visors anywhere in the world. Plastic-free visors have also been designed from paperboard and cellulose wood pulp. Covid-19 may initiate a shift back to local supply chains.

This is less certain, but countries may reconsider their dependence on international tourism. Tourism may recover fast, or it may be a long, slow recovery. The long-term impact on international aviation is equally uncertain. After all it was tourists, businesspeople and international students that enabled the virus to spread so rapidly across the world. It has certainly forced business and individuals to reconsider what is an essential journey. Consumers will need greater confidence before long-haul flights and cruise ship passenger numbers recover.

Covid-19, and enforced lockdown, has reminded us of the importance of community, family and our local environment.

It has shown that behavioural change at scale is possible and can be implemented quickly. The shift to home-working, and slashing business travel, has been dramatic and some changes will become permanent as employers can see the benefits. The case has been made to cut air pollution, and cities are reallocating street space to favour pedestrians and cyclists. Milan was the first to widen pavements and install more cycle lanes to enable social distancing and to reduce air pollution. Once in place, many of these schemes will become permanent and provide long-term benefits long after lockdown ended.

****

Covid-19 has led to major intervention in business by governments to protect jobs and the economy. In places these have been coupled with measures designed to improve the social responsibility of these businesses. France provided money to KLM/Air France conditional on a long-term decarbonisation strategy. They must cut carbon dioxide by 50% by 2024, source 2% of their fuel from renewable sources by 2025 and reduce carbon dioxide per passenger/km by 50% by 2030.

There is an opportunity to rebalance the economy, for example, switching expenditure on building new roads to investing in broadband and facilities to enable remote medical consultations. A priority should be to invest in projects with multiple benefits to improve our lives. Electric vehicle infrastructure will help to cut air pollution, and energy efficiency projects create jobs and make our homes more comfortable. Constructing district heat networks, strengthening the electricity grid and planting trees can all create jobs. There is also an opportunity to train the future designers and installers needed to convert our homes away from gas. Following the collapse in the oil price, it should be relatively painless to increase taxes on fossil fuels.

In a response to Covid-19, New Zealand announced a $1.1 billion package to support nature - to improve biodiversity, help tourism, protect agriculture and create 11,000 jobs. Money is being given to conservation agencies, regional environmental restoration projects and to control and eradicate invasive pests. This includes restoring wetlands, improving access for visitors to nature reserves, planting trees by rivers and $345 million to control invasive species – aquatic weeds, wallabies, possums and pine trees which can encroach on agricultural land. There is flexibility for these employees to return to their existing jobs when social distancing criteria are relaxed.

Covid-19 has come as a shock, but it is a foretaste of a bigger climate shock to come. The scale and impact of sea level rise and climate change will be much worse and permanent. Like the effects of climate change, in all countries Covid-19 disproportionately affects the vulnerable and impoverished. Covid-19 has shown how our politicians and societies are poor at assessing risk, and often sceptical of acting on the warnings from scientists. We could, and should, learn from this. We need to listen to scientists and act to pre-empt the worse impacts of climate change to come. There is an opportunity to appreciate and work with nature, to regulate our exploitation of livestock and wildlife, to transform our energy production and industry, and to implement behavioural changes that will improve our lives. We need international cooperation to do this effectively.

Conclusions

The inconvenient truth is that cheap fossil fuel has driven agricultural productivity, mass travel and access to cheap consumer goods. These activities are causing our climate to change. Much of this is due to excessive and extravagant consumption primarily by wealthy people. We all know and understand that our use of electricity, driving, flying and heating our homes drives carbon emissions. But the four 'hidden' elephants in the room are our excessive consumption of goods including fast fashion, our dietary demands including beef and dairy, society's use of cement and concrete, and the refrigerant gases and energy used for cooling.

We are failing. We really are not taking this seriously. And we have not yet seen what a changing climate and natural world will throw at us unless we change. Total emissions of all greenhouse gases have risen from 30 to 45 billion tonnes per year since the first global climate agreement signed in Kyoto in 1990 with half of all human emissions emitted from that date. We are perilously close to triggering the release of huge natural stores of carbon and methane from the permafrost and from forest fires creating an unstoppable spiral.

The scale of the challenge is enormous. We need to transform society in the next two decades. This will involve hundreds of minor changes and many big changes too. To make sense of all the actions proposed in this book, a **Green Action Plan** is included as an appendix that summarises the actions that government, business and individuals can take to address climate change. Governments need to set the policy framework and regulations; businesses need to provide us with better choices and individuals need to understand and take account of the impact of their choices on climate and nature.

We should restore nature to store much of the carbon that we have released from historical changes in land-use. But this is not enough; our priority must be to stop emitting billions of tonnes of carbon a year from burning fossil fuels. To stabilise our climate we will need to remove some of this carbon from the atmosphere and bury it underground. We need to understand the scale, interactions and limitations. There is insufficient land to sustainably grow the food that we need and to satisfy other demands such as biomass for heat, liquid fuel, construction materials and to manufacture bio-based chemicals and products. We must therefore prioritise based on sensible science.

Climate change is not an isolated environmental issue. It is entwined with population growth, poverty and inequality, habitat destruction and the pollution of our air, water and soil. We need to tackle all these alongside cutting our carbon emissions. For example, we can contain population growth through universal access to contraceptives and by providing social security to reduce the pressure to have lots of children to look after elderly relatives. And, easy access to clean water can free up time for young women's education, raising their aspirations which also leads to a fall in the birth rate.

****

The best time to plant a tree was 40 years ago; the second-best time is today. If I had a magic wand, I would travel back 40 years in a time machine and place a global carbon tax on the extraction of fossil fuel and ramp the tax up steadily over time. We could still set a global tax today, but unfortunately there seems to be no prospect of politicians agreeing to this. Instead, we need to design a myriad of complex policies to achieve the same ends.

The **solutions** could come from three sources:

1. A top down, formal process with governments cooperating through the United Nations, setting targets, policies and regulations.
2. Business pushed by shareholders and investors - influencing their supply chains and consumers.
3. Community and consumer choice - who we vote for, what we invest in, what we buy and how we influence one another.

The reality is that we need all three to work together but led by government regulations. For example, how can consumers make good choices if products and services do not have a carbon label on them? Governments need to regulate, business needs to apply the regulations and then retailers can offer consumers clearer carbon choices.

The common-sense solutions to our climate and nature crises can be summarised as:

- Reduce our need to travel and electrify remaining travel.
- Invest in innovative solutions to increase the efficiency of industry, decarbonise our electricity generation and heat production.
- Consume fewer manufactured goods, and for business to adopt circular economy approaches to what we do consume.
- Stop deforestation, restore degraded land.
- Shift to a predominately plant-based diet.

We need to believe and campaign for a future in which we respect and live in harmony with nature, consume what we need with no waste, enjoy satisfying lifestyles, live in strong communities - and all powered by renewable energy.

We need to direct and unleash the creative power of capitalism. Sensible government policy, aligned with innovation directed at sustainable solutions enables business to ramp up production, decrease costs and roll out solutions at scale - as evidenced by the dramatic fall in the cost of wind and solar power, and the rapidly falling cost of electric vehicles.

Tackling climate change is a **technological, economic, business, social and political** issue requiring **international** cooperation. Given their large and growing populations, what the governments and people of China, India and fast growing countries like Nigeria choose to do will determine if we win the battle. We need to develop and disperse the technology from advanced countries and to learn from the good example of countries like Costa Rica who are restoring their forests and improving lives.

If we could put a man on the moon in the 1960s, then we should be able to develop the technology that we need to solve climate change. We can tackle most issues with current technology, but further innovation results in better performance at a lower price which will make for an easier transition to a low carbon economy. The dramatic decline in the cost of solar energy has arisen from innovation and demand driven by government policies. This led to remarkable economies of scale in production. The same is happening with electric vehicles, led by China. It is likely that competition from plant-based substitutes and lab grown meat will fundamentally alter agriculture, particularly rearing livestock, over the next few decades. It is also possible that hydrogen made from renewable sources or technology to directly capture carbon dioxide from the atmosphere will have transformational impacts. If society places a sensible (high) price on carbon pollution, then this will spur innovation and introduce ideas that we have not even thought of today. The learning from Covid-19 is that a crisis can foster quick decision making, abrupt change and accelerate the pace of innovation. We can act if we choose to do so.

The momentum towards using renewable energy is unstoppable. As the demand for oil and gas falls, it will soon become uneconomic to explore and drill for new reserves. This will free up enormous sums of capital to invest in renewable energy. Economics drives the choices that we make. It is so obvious, it seems superfluous to say, yet the priority is to stop all subsidies for fossil fuels, synthetic fertiliser and other greenhouse gas emitting activities like cattle farming. Governments should phase out such subsidies, accompanied by support for those most affected. The next priority is to tax all emissions of greenhouse gases. This need not be an overall tax increase; we could reduce other taxes. These two actions alone could provide around half of the solution towards a carbon free world.

Companies and business organisations have a role to change the narrative around climate change from doom, gloom and complaint to a positive message of opportunity. Within companies the 'environment' needs to move from a health and safety compliance role, to a central position developing and marketing exciting and innovative new products. We need entrepreneurial, not incremental change. Energy efficiency, rooftop solar panels, electric cars, new heating systems, planting trees and environmental restoration will all create jobs. Social media influencers and people working in the creative industries; media, arts, culture, even marketing and sales; have an important role to shift attitudes and to promote low carbon lifestyles and products. Expertise from the declining oil and gas industry can transfer to offshore wind, producing hydrogen, geothermal drilling and carbon capture and storage. Business, and their trade bodies, should stop campaigning against new regulation. Instead they should actively promote sensible regulations to support the disruptive change that is necessary.

Polls suggest that the public are concerned about climate change, but do not always act on this. Sections of the public, often a vociferous minority, backed by a media storm often fight policies such as building wind farms or tax rises on fuel. It is therefore important to engage with the public to create solutions that are acceptable. The Citizens Assembly in the UK is a good example of government engaging with the public. Climate change is caused by what people choose to do. It is a people problem, yet most research is undertaken by scientists who are getting better and better at describing the problem and predicting its effects. We also need social scientists to focus on the solutions; some are technical but many more are societal and involve behaviour change. Social scientists can advise on the psychology of behaviour change.

Politicians are nervous about climate change as the next steps required are more radical than the relatively easy changes made to date. They are under little pressure from the public to act, feel constrained by the electorate, and receive unwelcome opposition if they suggest anything radical. Importantly net zero should not be viewed solely as a political left agenda concerned with regulations and workers' rights. Those from the political right also need to champion it through well-regulated markets and sensible government incentives. All governments should set the policy direction and introduce sensible policies, taxes and regulations.

For individuals in high income countries, the biggest changes we can choose to make are to buy an electric car, or even better give up owning a private car; avoid flying, especially long-haul; live in a smaller or shared home; refurbish our homes; stop frivolous and excessive consumption of goods; and eat a carefully selected, predominately plant-based diet. I have included a **vegan recipe** as an appendix as most of us are not aware of how we would shop and cook using plant-based ingredients. We also need to change attitudes through education and positive peer pressure. We need coal, oil and gas and air pollution to be the 'new smoking'. Soon society will view it as anti-social to continue burning fossil fuel. A starting point is a seismic shift in education, at school, university, work and adult education. Schools should teach children out of doors more; embed climate change across the curriculum and make ecology a compulsory subject at secondary school. We should also educate adults through publicity campaigns, education, and training at work or at night classes.

A world where we live in harmony with nature and are not causing climate change is surely a better place for us to live in. That is a **vision** that we can agree on. We can live in naturally warmer houses, breathe clean air and drink clean water, with less obesity and better health. Our neighbourhoods could be more pleasant, green places to live, where we can walk or cycle to access local services and work. Perhaps our quality of lives will improve if we stop obsessing about fashion, celebrities, our looks, consumption of goods and our social media image; focusing instead on families, our physical and mental health, education, satisfying jobs and close communities.

Living close to nature is good for our wellbeing. A shift towards a plant-based diet may free up land and provide an opportunity to rewild vast areas. But some aspects of restoring nature will cost money which carbon offset schemes can fund. Farmers and landowners will become stewards of the land, with obligations and subsidies to help nature. These benefits will cascade to improve the wider environment and our quality of lives.

I have probably upset most people at some point in this book - including research based academics, business leaders, farmers, fast-fashion retailers, heavy industry, aviation, the oil industry, pet owners and people who dislike government interference to name a few. Look at the bigger picture, think of your grandchildren and change what you can. You can then make further changes as government and businesses make it easier for us to make good carbon choices. Is this book radical? What would be truly radical would be to choose to deliberately and knowingly continue to fry our planet.

There is a climate and wildlife emergency. We all need to act now. Writing a strategy or setting a target for 2030 or beyond is not an appropriate response to tackle an emergency. It is in effect a self-congratulatory tactic that pretends that you have done something, when all you have done is delay action and leave it for someone in the future to act. There is little point in celebrating net zero achievements after our house has burnt down. The world will get to zero carbon emissions. The science and momentum are unstoppable. My fear is that we will get there a little too late.

People will ask if I am optimistic or pessimistic. But it is not about me, it is about you and everyone. Politicians, businesspeople, your neighbours, your friends, your work colleagues, people you will never meet. What Carbon Choices are they going to make? How are you going to influence them? What changes are you going to make? Your Planet Needs You.

If you liked this book please write a review on Amazon or Goodreads and share your thoughts with your friends, relatives and work colleagues. Face to face, by email, on social media. Together, we can make a difference.

If you would like to make any constructive comments to correct any errors or to improve Carbon Choices for a potential future edition then please email me at carbonchoices@gmail.com

Green Action Plan for Governments, Businesses and Individuals

This action plan focuses on wealthy, high income countries like the United Kingdom, but many actions will also be relevant to other countries and societies.

These actions are not comprehensive or a pick and mix selection. We need to do all of them, or equivalent strong actions.

For areas that do not fall within this green action plan, my five common-sense principles can help guide decision making:

1. Be fair across current and future generations
2. Price carbon pollution
3. Consume carefully, travel wisely
4. Embrace efficiency, avoid waste
5. Nurture nature

Government – Policy

- Accept that fossil fuels, travel, cement and industrial agriculture is causing a climate and a wildlife crisis and act accordingly.
- Set ambitious short-term carbon reduction targets, a medium-term net zero target and a net positive (regenerative) long-term target.
- Engage with the public and business to try and reach a consensus on the long-term strategic framework and policy direction of tax, subsidies and regulations.

- Engage cooperatively and constructively with other governments. Increase global ambition to cut emissions and boost nature. Support collaborative innovation and harmonise carbon taxes as far as possible. Provide funds to countries suffering most from climate change including a climate damages fund from aviation taxes.
- Shift the focus of university academic status from rewards based on pure research to kudos based on practical applications that benefit society and the environment.
- Invest in low carbon infrastructure that will improve our health and lifestyles such as electric charge points, cycle lanes and district heat. Invest to strengthen the electricity network.
- Advocate healthy, sustainable and low carbon diets to the public.
- Restrict building on greenfield sites, encourage regeneration on abandoned land, require zero carbon emissions, design in passive cooling, mandate solar panels and ban new gas connections.
- Improve the governance of water supplies. Place a sensible price on the use of water by industry and agriculture.
- Announce an early date to ban new petrol and diesel vehicles.
- Label all air freighted foods.
- Regulate the carbon offset market. Set an appropriate, and increasing, price for offsets. Establish a government backed permanent offset scheme for difficult to control carbon activities such as long-haul aviation.

Government – Taxes and Subsidies

- Abolish all subsidies for fossil fuel, fertiliser and other industrial and agricultural practices that emit any greenhouse gases.

- Shift the burden of taxes from income tax; to taxes on carbon, property and land ownership, extracting raw materials, road congestion and consumer expenditure (VAT).
- Set a cap, or tax, all emissions of greenhouse gases, including fossil fuels, fertiliser, ruminant livestock, industrial chemicals and cement. Set a higher tax on aviation to take account of its additional climate impact. Introduce a cruise ship levy. Raise these taxes over time.
- Incentivise renewable energy using market mechanisms.
- Levy VAT on new buildings and scrap it on energy efficient refurbishments.
- Tax or regulate products and imports with a high carbon or environmental impact such as rice, palm oil, aggregates, cement, batteries, metals and plastics.
- Offer incentives to share accommodation, for people to move closer to work, and to downsize to a smaller house more suited to their needs. Reform Stamp Duty.
- Set vehicle taxes correlated to their carbon emissions and develop an alternative means to tax private transport based on distance travelled.
- Create a programme and transition fund to help and retrain communities, business sectors and farmers who will be impacted by change to a low carbon economy.
- Shift marketing budgets to domestic tourism and create tax breaks for community-based tourism.

Government – Acting with Local Authorities

- Introduce strong land-use planning to encourage more compact, liveable and sustainable towns and cities. Reclaim our streets from car dominance and green our neighbourhoods. Restrict new building on greenfield sites.

- Promote walking and cycling. Encourage local schooling. Support car clubs.
- For existing buildings, require energy efficiency improvements, ban replacement gas boilers alongside a mass roll-out of alternative sources of heat street by street.
- Promote renewable energy through favourable planning rules.
- Stop building by the coast and start long-term planning for a managed coastal retreat.
- Teach ecology at school. Embed climate change and the effects of consumption across the school curriculum. Encourage outdoor education. Offer night classes or on-line learning to all of society.
- Run a massive and active campaign to provide healthy and low-carbon dietary advice, recipes and cookery classes in schools, workplaces and to the public.

Government – Farming, Land-use and Nature

- Abolish subsidies and tax fuel, fertiliser and emissions from ruminant livestock. Redirect the money raised to promote regenerative agricultural practices.
- Fund and create a much stronger network of inter-connected nature reserves.
- Restore wildlife rich habitats within national parks.
- License shooting estates and major landowners - impose a duty to restore ecosystems.
- Introduce tight regulations to prevent and tackle non-native species. Create a volunteer rapid reaction force to identify new non-native species and to control or eradicate existing invasive ones.
- Introduce tighter laws and more enforcement around the legal and illegal international trade in wild animals, eating bush-meat, wet markets and intensively reared livestock.

- Use money from carbon offsets to restore nature.

Government – Working with Business

- Mandate companies to 'do good for society' in return for continued legal privilege of limited liability.
- Require all companies to calculate their carbon footprint, identify measures to reduce it and to implement them over time.
- Work with companies and industry bodies to calculate the carbon footprint of food, products and services then introduce compulsory labels to inform consumers.
- Invest in near to market low carbon innovation and industry led challenge calls. Create innovation funds focused on difficult areas such as alternatives to cement and fertiliser production, energy storage, aviation, shipping and agriculture.
- Introduce tight, but practical regulations, to enable business to develop innovative solutions and encourage a circular economy.
- Make it compulsory for retailers to offer long product guarantees.
- Restrict marketing for goods that are harmful to the environment.
- Progressively tighten energy efficiency standards on all goods.
- Tighten the extended producer's responsibility on goods sold - all packaging to be composted or recycled, increased requirements for reused/recycled content in products.
- Tighten 'fit for purpose' consumer protection. Ban products where the adverse environmental impact is a greater burden to society than the usefulness of the product.

- Introduce a set of industry approved standard containers that can be collected after use, cleaned, and redistributed across industry and retailers for reuse.
- Pilot carbon capture and storage for heavy industry.

Businesses – Strategy

- Accept that fossil fuels, cement, travel and industrial agriculture is causing a climate and a wildlife crisis and act accordingly.
- Produce and market goods that you are proud of – those that are good for society and for the environment.
- Compete with your competitors on quality and price, collaborate on environmental issues.
- Adopt circular economy principles, including careful design of all new products to be durable, repairable and recyclable.
- Lobby government and industry bodies for ambitious, but practical and flexible, regulations to support a low carbon economy.
- Calculate your company's carbon footprint, set ambitious and ambitious short term targets and a net positive (regenerative) medium term target.
- Empower your employees with a mission to achieve sustainable solutions.
- If possible, reinvent your business model away from simply selling ever more quantity of stuff. Align profit with sustainable outcomes, perhaps refocus to provide a service. Maintain a long-term relationship with your customers and build brand loyalty. Compete on quality.
- Obtain an independent environmental certification. Work with international organisations that promote ethical and environmental good practice.

Businesses – Action

- Invest in all opportunities for energy and resource efficiency and on-site renewables.
- Reward bonuses to your directors and employees based on achieving ambitious environmental targets.
- Work with your supply chain to minimise the environmental impact of your purchases.
- Calculate and label the carbon lifecycle footprint of the products and services that you sell.
- Focus on innovation of new low carbon products, then market your products' sustainability credentials to your customers.
- Green your office or manufacturing site and encourage your employees to engage with community or overseas ecosystem restoration.
- Offset your remaining emissions in approved schemes.

Individuals – at Home

- Move home to near your work or child's school – walk or cycle to school.
- Buy or rent a property that is not too big or extravagant for your needs. Wherever possible, share living accommodation.
- If you are a homeowner, refurbish it and install solar panels. Adopt a low carbon heating system when incentives are available for your property.
- Wear clothes appropriate for the weather.
- Consider your digital footprint; stream on a mobile network and avoid listening to music with video if you can, select low definition video, switch off.
- Avoid waste, particularly food waste. Save water. Compost or recycle what remains.

Individuals – Shopping

- Buy food from wherever it is grown in naturally productive conditions. Avoid air freight.
- Restrain your consumption of status and frivolous goods – buy less stuff, campaign against fast fashion, buy small and energy efficient appliances. Buy second-hand goods and sell or donate goods that you no longer need.
- Avoid over-packaged and single-use products - particularly personal hygiene products.
- Rent or lease products and services from companies with strong sustainability credentials.
- Join a car club. If you need a private car then buy a small efficient electric model.

Individuals – other Actions

- Accept that fossil fuels, travel, cement and industrial agriculture are causing a climate and a wildlife crisis and act accordingly.
- Challenge governments, business organisations and individuals who seek to obstruct, delay or water down action to tackle climate change.
- Cut back on beef and dairy products. Shift to a predominately plant-based diet, and positively influence others' dietary choices for their health and the environment.
- Work for a company that does something socially useful; or retrain. Challenge your employer to 'do good' for society and the planet.
- Participate in local environmental community action – woodlands, wildlife, energy, sharing equipment, charities, educate and influence, tackle invasive species.
- Vote wisely; engage with politicians, decision makers, work colleagues, friends and neighbours. Be positive and constructive.

- Holiday wisely, avoid flying, or value the occasional flights that you take.
- Choose sports, hobbies, art and culture that you can view or participate in locally.
- Consider ethical banking and investments. Pressure your bank to stop investing in fossil fuels.
- Pay to offset your emissions (accepting that this will not solve climate change).

A vegan recipe

Sweet Potato and Lentil Soup (serves 4)

Ingredients
- 2 tablespoons olive oil
- 2 onions, chopped
- 800g sweet potatoes, peeled and chopped (2 large or 4 small)
- optional carrots or parsnips
- 2 crushed garlic cloves
- 2 teaspoons coriander
- 2 teaspoons curry powder or chilli flakes
- 1.5 litres of vegetable stock (2 stock cubes)
- 200g red lentils, rinsed

1. Put the olive oil and onion in a large saucepan and fry for 4 minutes until softened.
2. Add the sweet potato (and optional carrots and parsnips) and fry on a high heat for 4 minutes. Add garlic, coriander and curry powder.
3. Add the stock, lentils and salt and pepper to taste.
4. Simmer for 20 minutes. Add more boiling water if necessary.
5. Blend the soup to thin it.
6. Serve with crusty bread.

Mexican Butternut Quesadilla (serves 4)

Ingredients
- 1 large butternut squash, peeled and chopped (or use pumpkin in autumn)
- 1 tablespoon olive oil
- 4 spring onions, finely sliced
- 3 crushed garlic cloves
- 800g kidney beans (tinned) or equivalent
- 1 tablespoon BBQ seasoning (or mix of alternatives such as paprika, chilli, cajun)
- juice of 1 lime
- 8 flour tortillas

1. Place the butternut squash in a large microwaveable bowl with 2 tablespoons of water. Cook on high for 12 minutes or until soft.
2. Add the olive oil to a frying pan, fry the spring onion and garlic for one minute. Add the beans and BBQ seasoning to the pan and fry for two minutes.
3. Tip this mixture into the butternut squash bowl and add the lime juice. Mash it all together.
4. Put the frying pan on high heat. Add more olive oil if necessary. Place one tortilla in the frying pan and fry for one minute. Pick up the tortilla, place another it its place and pile one quarter of the mixture on top. Place the first tortilla on top to make a sandwich. Fry for two minutes until golden brown.
5. Repeat with the other tortillas (Tip – you may keep the completed tortillas under a grill to keep them warm whilst the others are frying).
6. Serve with any salad (such as a mango salsa).

Mango Salsa

Ingredients
- 350g mango, chopped (1 mango)
- 150g cherry tomatoes, chopped
- 80g onion, chopped finely (1 onion)
- 100g red bell pepper, chopped (1 pepper)
- 1 crushed garlic clove
- juice of 1 lime

1. Stir mango, tomatoes, onion, red bell pepper and garlic in a bowl. Add lime juice and season with salt.
2. Let sit for 5 minutes for flavours to blend. Store leftovers in refrigerator.
3. Serve with tacos as a starter or with Butternut Quesadilla (above).

Selected References

A small selection of books and references that inspired or influenced Carbon Choices.

Books

Feral: Rewilding the Land, Sea and Human Life, George Monbiot, 2014

Half Earth: Our Planet's Fight for Life, Edward O. Wilson, 2016

How Bad are Bananas, Mike Berners-Lee, 2010

Scotland: A Rewilding Journey, Susan Wright, Peter Cairns and Nick Underdown, 2020

Sustainable Energy Without the Hot Air, David MacKay, 2009

The Uninhabitable Earth, David Wallace-Wells, 2019

There is no Planet B, Mike Berners-Lee, 2019

Wilding: The Return of Nature to a British Farm. Isabella Tree, 2019

Reports (available on the internet)

Environmental Impacts of Food and Agriculture, Hannah Ritchie and Max Roser, 2020

Global Assessment Report on Biodiversity and Ecosystem Services, 2019

Carbon Choices

IPCC Special Report: Global Warming of 1.5°C, 2018

Our World in Data – University of Oxford based website (includes information on food carbon footprint)

News related Websites

Business Green
Carbon Brief
ECIU: Energy and Climate Intelligence Unit
Edie Energy
Guardian newspaper
WWT Online
Yale Environment 360

Printed in Great Britain
by Amazon